Touching Animal Souls

Published by: Kima Global Publishers

Kima Global House,

50, Clovelly Road,

Clovelly 7975

P.O. Box 22404,

Fish Hock 7974

South Africa

© Gabrielle Harris 2010

ISBN: 978-0-9869858-3-6

e-mail: info@kimaglobal.co.za

Website: http://www.kimaglobal.co.za

Cover design by Nadine May

All rights reserved. With the exception of small passages quoted for review purposes, no portion of this work may be reproduced, translated, adapted, stored in a retrieval system, or transmitted in any form or through any means including electronic, mechanical, photocopying or otherwise without the written permission of the publisher.

TOUCHING ANIMAL SOULS

Developing Awareness Through the Animal World

By Gabrielle Harris

There is a great knowing inside us all that is borne out of a universal true intelligence. It is always there, staring us in the face, inviting us to feel its perfection.

Dedication

This book is dedicated to my mother, Clare, for knowing and encouraging my passion, and to my father John Luyt, who inspired it. Words cannot express my gratitude. This book honours, with tenderness, the many animals that have touched my soul. Too many to mention.

Acknowledgement

My heartfelt thanks to so many souls. Julie, Johnny and Robyn for being there. My world has been expanded so much because of the angels of patience, peace and perception – my husband Darryl, and sons Zac and Kai.

To all the trainers who have been a part of my learning experience. Sea World trainers – many of them, past and present. To name a few – Charmaine, Hazel, Susan, Sue, Lee, Tracey, Ryan, Carey, Alli, Zama, Reggie, Martin, Judy, Debbie, Kelly, Sarah, Colette, Hayley(s), Margaret, Sean, Kerry, Arenea, Terry, Mhlaba, Lungi, and many more. Trainers from facilities around the world who have taught me, inspired me and even let me sleep on their couches during wonderful travels - Wayne Nicol, Brian Ma, Chris and Morgane Davis, Ken Ramirez, Frantisek Susta, Radek Zoula, Linda Clockie, Dave Leslie, Patsy Divine, Kim Terril, Mr Kwek, Dr Alaa, Mohammed and Mohammed and Mohammed, Mr Fathay, Leanne Jamieson, Deon Venter, Alta Viljoen, Geraldine La Cave and the list goes on.

Facilities, organisations and friends who have enabled my growth and experiences. Stephen Norval from truePoint Foundation, yoga guru Jean Arundel, Shaun Knock, Tony de Freitas, Judy Mann, Rudy van der Elst, Mark Penning, Wynter Worsthorne, Dave Morgan , Adrian Tordiffe, Robynn Ingle Moller, John Werth, Dominic Moss, Jimmy Amany, SAAMBR, IMATA, the ABMA, PAAZAB, the AKAA, Horse Gentlers International, South African Police Services, the ARK, Prague Zoo, Sea World Orlando, National Zoological Gardens(SA), Disney Animal Kingdom, Shedd Aquarium, Brookfield Zoo, Hawaii Dolphin Quest, Hawaii Sea Life Park, Boudewijin Park, Park Asterix, Duisberg Zoo, ZooMarine Portugal, Lisbon Zoo, Giza Zoo staff, Singapore Underwater World staff, Cango Safari Park, Two Oceans Aquarium, and more.

Photographers who have allowed me to include their pictures. Peter Pinnock, Maitland Simms, Darrol Williams,

Touching Animal Souls

Jane Smith, Gareth Malyon, Bianca Lee-Hawood, Kelly de Klerk, Darryl Harris, and Leanne Jamieson.

These are just a few of many people who have enriched my life. There are many more. Truth is, if I have interfaced with you, you have left me richer. These are human names. If I had to write down animal names, it would take a chapter. Many of the animals are named in the book. There are many more. Thank God for the animal kingdom. Words don't cover it. To name a few, Gambit, Gandalph, Mushatu, Frodo, Jula, Khwezi, Kelpie, Kani, Freya(s), Kela, Sandy, Seamus, Whiskey, Clementine, Ingelosi, Tombi, Khanya, Zulu, Affrika, Vic, Benji, Shadow, Gimli, Boom Boom, Sasha(s), Pepper, Teabag, Dee, Mogwai, Prince, Peter, Gremlin, Wednesday, Morticia, Donald, Rum, Cola, Mahala, Black Tie, Ice, Khethiwe, Jabu, Illanga, Thuli, Bowie, Hlabathi, Moya, Thembeka, Dougal, Gerry, Terminator, Jaxon, Hobbit(s), Peanut, Po, Pip, Pavarotti, Chocolate, Hagar, Scratch, Boytjie, Buddy, Savannah, Cat, Max, Hermione, Philli, Tina, Cleo, Whiskey... The list is endless. I could go on and on...

Julie, thanks for assisting with the editing process. My publishers, Robin and Nadine.

Table of Contents

Preface 12

Introduction 15

CHAPTER 1 *Leadership – Giving and Receiving* 26

Trying or Being 26
Proactive relationship building 29
Leaders with whom We wish to be Associated 31
Pecking Order Misconceptions 36
Assertive, Aggressive or Submissive 40
Playing is being Present 42
Personal Belief Systems 43
Listening 45
Leadership vs Dominance 47

CHAPTER 2 *Developing Leadership qualities in Ourselves* 51

Pitiful Confusion 53
Maternal Empathy 56
Mean Yes When You Say Yes 58
Be Clear that the Gift is Deserving 61
A Good Leader is a Reward in Action 63
Simple is Effective 63
Justice will be done, so be Fair 65
The Role of Body Language and Posture 66
We Get What we Give 71
Grey is Confusing, Black or White is clear 73
Little Tricks of the Trade 73
Feel It 75
Quality of your Attention 75
Taking Responsibility for our Actions 77
Play Play Play 78

CHAPTER 3 *Sensorial Animal Oneness* 80

Awareness Alone 82
Training Trainers to be Confident 83
Managing the Dynamic of a Group 84
Trust in the Relationship 85

CHAPTER 4 *Why we do What we Do – Expressing Who We Really Are* 87

First Lesson 88
Loving them enough to lose them 89
Purpose 90

CHAPTER 5 *The Power of Choice – Free Will* 93

Its all in your head 94
It's not personal 97
Emotional Consciousness 99
Choice is power 99
Who are we? 102

CHAPTER 6 *Responsibility – Cause and Effect* 104

No problems!! 104
Anthropomorphism 106
Power Talk 107
Belief systems and excuses 109

CHAPTER 7 *It's not always as it Seems: Beliefs and Illusions* 112

Voice in your head 112
Alpha and Beta mind states 113
Alpha Spiritual State 114
Confidence and the Alpha State 116
Underestimating Animals 119

CHAPTER 8 *Intuition and Feeling* 121

What are the Limits of Animal Communication? 121
Intuition and animal care 125

Table of Contents

So do You have to be a Witch or a Wizard to do this? 126
Where are the Edges of Our Experience? 133
Intuition Improving Communication 137

CHAPTER 9 *Developing your Potential to be Intuitive* 140

First Impressions.. 141
Stuck in the Mind 142
Feeling the Present Moment 143
Is Your Canvas Coloured or Ready for a New Picture? 146
Acknowledge and Exercise 147
Again – Never Stop Playing 150
Feedback 151
He Who Hesitates is Lost................................... 152

CHAPTER 10 *Comfort and Discomfort* 154

Learned Helplessness....................................... 154
Fear versus Love 155
Discomfort is a Message 159
Be Open 160
Identify Fears 162
Stay at your Pace... 163
Watch Closely 163
Confidence Breeds Confidence............................. 164
In the Beginning 165
Sing out to be Consistent 167
Wishy Washy is Diluted Nothingness...................... 168

CHAPTER 11 *Listening to the Signs – Mutual Support* 169

Simple Feedback 170
No Wrong or Right, Just 'What Is' 171
Tolerating Abuse... 172
Objective Interface 174
SLAM Principle... 176

CHAPTER 12 *They Reflect our Inner State –*
The Mirror Principle 178

Reflecting our Anxiety 179
Children and Confidence 181
Heaviness Rubs Off on Them 183
Animals as Psychics 184

CHAPTER 13 *Being Clear – Non-Judgement* 186

Doing what needs doing 187
Judge Jury & Executioner – in our Head 189
Language Kills the Moment 193
Moya's lesson saves lives 193
Limiting Potential with Judgement 194

CHAPTER 14 *Intention – The Universe*
Handles the Details 197

Just Be 197
Visualise End Result 199
Visualise and the Universe will Handle the Details 201

CHAPTER 15 *What we Focus on Expands* 204

Our Misguided Focus Directing their Action 206
When to redirect 208
Reward only what you want 209
Entrenching Behaviour 210
Extinguish unwanted behaviour 211

CHAPTER 16 *Ego – Friend or Foe* 214

What is being Present all about? 216
What is it that causes us to lose the moment? 216
Subject Consciousness 220
Object Consciousness 221
Leadership vs Dominance – Which Style is Self-Aware? 221
Inter-Human Ego 223
What can we do to Overcome and Diminish Ego? 224

Table of Contents

Don't get Lost in the Bridge –Go with the Flow 227
Don't take their natural inclinations personally 228
Animal Training is a Lesson in Self Awareness 230
The Bottom Line 234

Conclusion 235

Appendix 239

Bibliography 269

About the author 270

Preface

It may not be a coincidence that the word 'KNOW' includes the word 'NOW'. Because when we are in the now we are not prejudiced by beliefs, past or present. We are simply experiencing what is, and responding as an animal would – with truth and clarity. It is when we truly know.

Human beings make a difference on this planet. Each of us has a responsibility to determine what difference we choose to make. It is possible that the sadness that is found on our planet is simply a reflection of the sadness, unconsciousness and fear that is imminent in the fabric of most of humankind. This is what I find myself believing more and more each day. In spaces where I recognise joy and positive energy in human beings, I always note that the natural world reflects that energy.

To have the good fortune of being in relationship with animals is an extraordinary privilege. When we look closely at that relationship it is possible that we find an unconditional being facing us, one that is there with the sole purpose of leading us to a more conscious existence. Imagine that. Imagine the possibility that our every action and reaction affects every aspect of the world. There is no wrong or right way to be

Preface

relationship with that concept, except to say that if we accept it as a belief system we need to take full responsibility for all that occurs around us. When we take responsibility, then we will affect the change that is required that will ensure that our planet becomes inheritable to our children.

The truth is that animals can lead us to this end. As an animal trainer, this is what I have learned from them. They are ever-present, watching intently and completely aware of all that is in their environment at any moment. My gurus. My spiritual guides. They are not worrying about what to wear tomorrow or how to clean up the mess they made yesterday. They are in the now. They survive in the best way they know how. That is not to say that they do not learn. I train animals, and know that they can learn the most remarkable lessons. They are however unconditional in their responses. No hidden agendas. Just acting to survive. And generally, when in relationship with humans they are reacting. If we listen and watch really closely we will be gifted with an astute picture of ourselves, and if we receive this picture, we can work towards being more responsible planet dwellers. Darwin's theory of evolution has been interpreted by many to place man at the top end of the pyramid. In other words, placing him above the animal kingdom. Perhaps the reverse is true. If we are searching for pure consciousness, the goal of all

faith and religions, then we need to aim towards the simple life forms. Perhaps the amoeba is the ultimate life form.

This book was written on a few levels. On one level I have shared many anecdotes that I have experienced in my time with animals. I have been enriched by these experiences and trust you will see their colour and excitement. On another level I elaborate these anecdotes to share life lessons that I have learned in these interactions. Finally, one of these very large lessons is that trainers who are more humble and open to learning the lessons that animals are here to teach us, are more effective animal trainers. So, if you are an animal trainer or care taker of pets, the insights shared in this book have the potential to improve your abilities to relate to animals. I would love to know your stories and hear your thoughts on the ideas presented to you in 'Touching Animal Souls'. So, keep in touch, and happy reading.

Introduction

Angels come in all shapes and sizes

Nervously I took hold of Gandalph's halter and lead him out into the field. We both stepped hesitantly, me worried about his reaction, and him probably not trusting this hesitant person at his side, maybe sensing a predator in my anxious pose. A couple of steps out of the smaller arena and my heart was in my throat. The young colt probably felt the same way. It was as though we were stepping into the unknown, and all at once within moments of us leaving the confines of the fenced area, the young horse bolted. Although only six months old and not nearly fully grown, he took off with intent and I was not able to hold onto his lead rein. I tripped and fell and his hoof clipped the side of my head. My pride was knocked out of me in that instant. I stood up, stunned, and felt a damp trickle of blood on the side of my face. Gandalph had circled around and was standing a few feet away from me. With adrenalin filled bravery I went up to him, firmly grasped his halter once again, and this time looked into his beautiful blue Cremola eyes, and in a sudden sense of humility, saw a teacher there. Out loud and without hesitating, I remarked, "Okay sir. I am ready to learn the lesson."

Touching Animal Souls

I have no clue where those words came from, but the moment did lead to a whole new direction in my career as an animal trainer, and a new road in the personal journey of my soul. The moment heralded yet another animal inspired humility driven look inward into who I really am as a person. I lead him back into the smaller paddock and the lesson began. The teachings that Gandalph inspired that day have panned out into my own careful investigation into me. Because, what I have recognised is that everything that is going on in me is manifesting in my environment. And when I am working with animals, the manifestation is always a pure reflection of my inner self.

Animals are incredible souls, and largely, they do what is required of them. They don't take things personally. They just are, being in their environment, accepting what is and doing what needs doing. Their environment is their universe, and their universe is their guide. And if we are observant, we will notice that this environment is made up of everything including ourselves. The result is that in relationship with animals, we are looking into a powerfully insightful and wonderful mirror. A mirror that in fact reflects both ways, if we let it. Because it is operating between the souls of humans and animals. With animals before us, there is never a need for the psychologist's couch. They ask all the questions and point out all our cracks. True natural therapy. If we care to listen.

When Gandalph kicked me in the head and woke me up, I had been formally training animals for years. I had not had much experience with horses however. I remember riding horses as a child. I was around eight or nine at the time. My childhood riding instructor was a young man, whom in retrospect, I realise did not have much faith in my ability to ride. I was always given the slowest plodder to mount, and I remember being scared. I was also desperate to keep up, and my plodder would never listen to my urging; probably completely desensitized to the constant nagging of unconfident young lasses like myself. But I was desperate to develop a relationship with horses. Just like

Introduction

many young girls, I had books on the subject, and fell asleep at night with "My friend Flicka", "Royal Velvet" and "Black Beauty" dreams in my head, while staring at posters of handsome stallions pinned to my bedroom walls. Unfortunately, my riding lessons only really developed a sense in me that I was not good enough. I fell off a lot. I recall a lesson where a horse was being 'obstinate', according to the instructor. He became annoyed and asked the student who was riding the horse to jump off. My feisty instructor jumped on the horse and began beating the animal. The horse reared and bucked, and the instructor shouted louder and louder. I was terrified. At the time, I was not sure what was making me afraid. I think now, when I feel those feelings of fear around animals, that the chance exists, that I am recognising empathetically, the fear that the animal is feeling. Not long after my instructor's display of bravado, I gave up on the riding lessons. My confidence was dashed, and discomfort had set in at those weekly nerve wracking expeditions. It just did not feel right. It became a shelved fantasy – for the time being.

My father is a vet. My growing years were spent at his side, watching what he did with awe and wonderment. I remember him having to handle really difficult animals, dogs that were incredibly aggressive and frightened. He would gently take charge of the situation and manage to do his job successfully, because of the discerning confidence that is his demeanour. Our home always had an orphan animal or two in our midst. Some were wild animals that were released back into the bush after they were raised or rehabilitated. Others were pets that became a part of our family having been left in my father's surgery for one reason or another. My dad's surgery was on the bottom floor of our home, so the stray and abandoned animals only had a short distance to travel to find a new home.

There was a three legged Siamese cat called Cat. He had been caught in a snare and after the necessary amputation of his front leg, his owners did not want him back, so he was brought upstairs. Lady Button, a Maltese cross who was brought in to be put to sleep because her owners were moving. Along with four puppies. We found homes for the puppies, and then Lady bounded up stairs to become a lively part of our family. Jasmine the kitten that was hand reared after she was found singed and close to death in a factory fire, Nicolas the hedgehog who was displaced by suburban development. Rehabilitated under the cupboard in the lounge and then released back into the bush. Fluff the mouse bird who fell out of his nest as a chick and ended up in a sanctuary after we taught him to fly. Just some of the teachers I had when I was growing up – some of the amazing inspiration I had the pleasure of knowing. Even then I took the relationships for granted, and did not even know I was being taught. It is only with hindsight, that I see that I was the student. And at the time all I thought was that I was the loving human who was there to save them; the hero filled with self importance, and devoid of the necessary humility.

My rebellious teenage years left me without a clear picture of what I wanted to do with my life. Sub-consciously I probably always knew I would be involved with animals, but I had not mapped out a clear path to my destination. I went to University and in my bachelor's degree, did some psychology and Speech and Drama. This was a good move on the part of my sub-conscious, as it would hold me in good stead for my job offer in the world of animals. I had no thought of that at the time, however. I spent some time travelling before I went to university, and my goal when I was studying was to continue with the travelling after I achieved my degree.

To save for my trip overseas, when I finished studying, I took on many different casual and part time jobs. These included everything from aerobics instructor to waitress. At some point in this mayhem of activity, a friend of mine called and said there was a job going at her place of work, and invited me to try out in an interview. She worked at Sea World

Introduction

in Durban, in the mammal and bird section with the dolphins, seals and penguins. I had always been curious about what she did for a living, but had never imagined the possibility of doing it myself. With nothing to lose, I agreed to go for the interview. I considered that it would be interview experience. The job in question was for a marine animal apprentice trainer and show presenter, and because of my bachelor degree, I was a good potential fit for the work. In a whirlwind interview process, and before I had time to carefully consider my future, I accepted an offer of employment and my plans for travel changed for good. I was hooked.

I was introduced to the dolphins and in the process one of them rolled over onto its back so that we would tickle his stomach. The manager who was showing me around laughed and told me to look at his belly button. I was amazed. These animals from the sea have belly buttons. At once they became individual personalities, each with their own character and story. And that was the start.

Gradually I became more and more embroiled in the world of animal care and training. I was bestowed the honour of making friends with some incredible dolphins, seals and penguins, and learned the art of training animals using operant conditioning and positive reinforcement. I became fairly confident in the methodology, and felt sure that I was working in an equal partnership with the animals I trained. With success I used my new found skills on dogs, rats, cats, pigs and an assortment of other creatures with which I came in contact. I even adapted it for use on my children when they were born.

By the time I met Gandalph, I was well versed in the technology, and quite frankly believed I knew most of what there was to know about animal behaviour modification. I had even achieved further academic qualifications on the topic. This self-importance was shattered with Gandalph's kick. This know-it-all was off track with all that pride. It was as though the animal kingdom was laughing hard, and as a joke, sent me

Gandalph. I needed humility to continue my journey into the world of relating to animals. And what was required for me to achieve that humility was for me to be toppled off my holier-than-though pedestal, or more rightly, be taken out of my head and placed into my heart. Gandalph taught me the biggest lesson I have learned so far, not only in my career, but in my life. So that I would listen to the lessons, he literally kicked my pride out of me, leaving space for that humility. He showed me that listening is as important, if not more so, than talking.

Gandalph taught me that for me to progress in my knowledge of animal behaviour and behaviour modification, and even more than that, for harmony in my life, I needed to look within. I needed to recognise that for my relationship with animals to work effectively, I needed to be an effective leader. This is not to say I am in charge of the animal, but more importantly, I need to take full responsibility for my actions. For me, assuming the role of leader is not an egotistical step. It is simply a case of being fully conscious and deliberate in my actions. In fact, in my life, and in relationships with everyone, when I assume this responsible attitude, I have consistent success in my dealings with those around me. To be clear here, I see an enormous difference between leader and dominator. This lesson will be elaborated on in further chapters. But at this point suffice it to say that taking charge of my life and accepting responsibility for everything that happens to me is the exciting result. So, I don't get to blame animals or people around me for anything. It is always my doing.

Animal trainers love to share the behaviours they have trained animals. Some amazing things have been taught too. At conferences, trainers share the most extraordinary accomplishments. Training dolphins to lie still in the water for voluntary artificial insemination procedures. Teaching rhinos to stand still while their tusks are filed down by the vet. Getting male lions to urinate on cue. Incredible. Trainers are credited as a result of their successes. The truth is that the behaviour is

Introduction

not the reason for the trainer's success. The groundwork is the essence of it all. Many people have relationships with animals and do not formally train them. This is another lesson that Gandalph has imparted. It is all about relationship. If I am to be effective in my role as animal trainer and animal care specialist, and human being, I need to conduct myself with the end in mind, understanding that whatever I do affects the total well-being of the animal in question. And the end is the relationship. Perhaps 'end' is the wrong word. Foundation is probably a better way of describing what the fundamental goal of relationship is all about. All relationships are a result of some kind of communication. When training animals, any behaviour I train is only a by-product of that relationship. Without relationship there is no fulfilment in our interactions. What I learn as an animal trainer teaches me a great deal about myself. This learning is a continual lesson and one of the reasons I find it really fascinating to be in the profession I have chosen. Getting to smile at how I distort my reality after being made aware of it by the creatures with which I interface. Funny that I find it easier to take their advice than the advice of people. Their advice is always unconditional.

I mention clarity in communication. To be clear, our feelings and intellect need to be harmonious. I was brought up as a Catholic. It was a strict religious upbringing, with lessons and instruction that guided my developing mind. It made sense, on a theoretical level and inspired a God-fearing intellect. However, I never once, in my religious instruction and education at a catholic school, felt close to God. The catholic manner of interacting with God was taught to me as a very intellectual lesson. I liken this to the pure theory that is taught to animal trainers. It is only a part of the equation. The first time I ever had a catholic spiritual experience was when I visited the tomb site of Vasco de Gama in Portugal. I was totally blown away. It is in a beautiful ornamental setting, but what really made me feel that way was when I saw that the Poet that had accompanied him on all his adventurous cruises on board his ships, was given as honourable a burial site as the historic Captain. I suddenly recognised that there was warm hearted humanity in the history lessons I had been instructed on in my school classrooms. For the first time that history felt real. It was the first time I found out that all those cruises were documented in poetry.

The fact that art had such credibility in those early times excited my wild side. Art is the way that human beings have expressed feelings throughout history. It is my experience that when people are inspired by feeling, they are able to act to achieve marvellous things. In this book, I will relate how I have found a need to achieve a measure of synchronicity between feelings and intellect in order that I maintain successful relationships with animals. Another life lesson. Feelings are the way in which we experience life. Without them we are flat lining. It is possible and even necessary to train animals using feelings. The synchronicity between feeling and intellect is something that is required in all of life, but once again, the lesson for me is one I have solidified in relationship with the beasts of this world. I had success training animals, but it was only when I learned that feelings were a very necessary part of my interactions with the animals that my communication with them became clear enough. And this improved my ability as a trainer.

Working with animals is an inspirational activity. A Doctor Doolittle experience. We are communicating with them. Responding to them and having them respond to us. Just as all successful relationships with humans demand that we take total responsibility for our own actions, so too do relationships with animals. Whether the animals are domestic or exotic, they are sentient beings, capable of feeling, and very affected by our presence. I have heard the arguments that animals don't have feelings, that they are not thinking in the first person, but I do not buy these arguments. I have seen too many animals solicit communication, in an effort to tell us something; often something very important. The majority of

Introduction

my experience is training animals using operant conditioning with reinforcement. For anyone who is not familiar with this methodology and interested in learning more, I have outlined the basic principles of this method and my opinions of the various tools of training in an appendix section to this book. However, this is a very brief outline as there is a great deal written on this subject already by people far more qualified.

The objective of this book is more than outlining behavioural modification principles. I will recount the lessons the animals have taught me. And if I understand them correctly, it is that to clearly and successfully interface with animals, it is important that we take the time to objectively assess where we are coming from on a personal level. It is possible to communicate on many different levels. Training animals is theoretically based. There are definitions and descriptions of any move we can make. However, and this is the message for me, I am the one implementing the definitions and techniques. In life society tells us that there is a right and wrong way. But it is only our experience, and our own personal ethical discernment that leaves us with a feeling of being 'right' or 'wrong'. Our belief systems are our subjective mask that colours things personal. I know that what I am feeling will affect how I interact with and sense the world – and this will affect how the world reacts to me.

In my experience as an animal trainer, and trainer of other animal trainers, I have found that I need to take careful cognisance of what I am communicating. In order to do that effectively, I need to be very conscious of my state of being at all times. For this reason, as an animal trainer, it is my imperative to be on a permanent mission of self discovery; to achieve consciousness as it were. This was always an airy-fairy concept for me when I heard people talk of it in spiritual circles, but consciousness is actually very simple. It is just being present with how I am feeling at any give time. This, as will be outlined, is essential when I am training animals, because, if I am in a bad mood, for example, I will communicate very differently when I am training than if I was calm. On this journey, I have become increasingly aware of the fact that when I am clear in my own right, I am better placed to interact favourably and communicate productively with animals and people with whom I am in relationship.

I have had the good fortune to learn many humbling and fascinating lessons and I know with every measure of my being that the journey is not over. There is still much to learn. I get the sense often that the animal kingdom is still smirking, and enjoy their sense of humour. Gandalph has taught me to laugh with them. You cannot do this, or life without the necessary sense of humour. Humour is the window to humility. The lessons I have learned improve my abilities, but more than that, I have found that my less formal relationships with my personal menagerie of dogs, cats, horses, birds, snakes and a pig at home have also improved significantly. I am less reactive with my animals at home, and a great deal more receptive to what they are reflecting in me. Finally, and most interesting is that the lessons have also significantly improved my relations with people in my life. To be sure, I have not reached the end of the road. I look forward to more adventures with the creatures who have taught me so very much more than I have ever taught them. Lessons I would like to share with all of you.

If we are confident and clear and unconditional with ourselves, the same will be true for our relationships with animals. We need to be able to discriminate between our egos and our true selves. At the heart of our being, beneath our egos, lies limitless possibilities that are available to us. In situations where I have had the good sense to put down my fears, preconceptions and thoughts on how things 'should or could or would be - if only', I have felt liberated to be present and free to do anything. I see my ego as a wonderful messenger, reminding me that there is a bigger picture to consider. It is helpful for me to recognise my ego as that little voice in my head telling me what to do, judging me and others and

Introduction

attempting to rule my existence. If I recognise that there is someone listening to that ego, then I realise that I have the power to choose to listen to the voice or not. If I take a step further into recognising the discrimination between the voice and my observer, I recognise that all my fears and concerns and conditioned attitudes are manifesting in that voice. They are limiting the way I experience the world and limiting my ability to act appropriately. The observer to this little voice is much wiser, and unlimited with potential to accomplish so much more than I can ever imagine.

Chapter 1

Leadership – Giving and Receiving

Trying or Being

I have an incredible yoga teacher. She delivers pearls of wisdom while tying us up in knots during her classes. After a pretty intense section in one of her classes where I had just exercised all my stamina and the extent of my strength, she invited us to lie still for a moment and feel what we had just done. As we lay there she shared *'When we are done with the doing, which is the giving, always have time to just be, to receive.'* I had never associated the act of being present – to be – with receiving. I was emotional hearing this remarkable statement. I interpreted that being conscious and present required absolutely no action. Like many women and probably men too, I can steel myself against insults, but I don't receive compliments or gifts very gracefully. A compliment will bring me to tears where an insult will leave me cold. In order to be conscious and clear, which is the spiritual yearning of so many of us, I need to be in a state of receptivity. I need to be able to receive without judgement or anxiety.

Chapter 1 Leadership – Giving and Receiving

This lesson has great significance to me as an animal trainer. Basically, trying to do something with an animal will never work because when I am **'trying'** I am not in a state of receptivity. The act of trying is conditional. The end is the objective. I will then actually be dominating the animal. When on the other hand I am present and unattached to the end goal, I will be receptive to the animal's communication, and thus more inclined to work with the animal towards a goal. This is leading the animal as opposed to telling it what to do. Thus, to train animals, some doing and being is required. Giving and receiving. To be completely clear in this, the giving actually becomes the receiving.

Frodo is a fantastic matriarch dolphin. She is an incredible mother and grandmother. She was one of the first dolphins I ever trained. I still play with her today, and I never have a training session with her where I don't get to smile at her wise and challenging manner. Challenging in the nicest possible way. When I am with Frodo I truly feel like I am lying on a couch in a very clever psychologist's consulting room. The therapist is asking me one question, over and over. The question is "Are you sure?" Frodo will only be in relationship with me, co-operating and enjoying herself, if I am totally focussed on the session. So, it was no surprise, when early in my career I was working in a show for a bunch of visiting friends which included a then potential beau, and Frodo was refusing to do anything. She had been perfect all day and all of the week before in all shows, so I could not imagine why she was choosing not to perform. I was feeling a little shy in the show as a result. She was swimming on the opposite side of the pool, and not interested. The presenter was 'filling'. This is what we have to do to make the show go on. When the presenter begins to 'fill', it usually looks as though it is a part of the show. They will chat about dolphin facts and such subjects. But as they continue to fill, they usually resort to actually saying something to the audience along the lines of 'these are not animatronics. We cannot wind them up and set them off to do a show. Sometimes they just don't feel like doing it.'

This slowly brings the audience's attention to the fact that not everything is going as planned. To make light of the situation this particular presenter actually asked me, on the microphone, if I had put on the wrong deodorant that morning. This did very little to boost my confidence. I simply could not stop imagining my potential boyfriend's smirk. I wanted the world to swallow me up whole. I was making no progress standing at the pool side, so stepped back and pondered the situation. For probably the first time that session, I looked at Frodo properly. I watched her and noticed that she was watching me. She obviously saw that my focus had changed. So hers did too. She hesitantly approached my side of the pool and looked at me out of one of her eyes as she lolled on the surface with one pectoral fin breaking the surface of the water. It was as though she was summing me up. I forgot about my social life crumbling, and simply smiled at myself. I remembered why I adored this lady in front of me. And I was in session with her again. She had asked the question again, and I was sure. I stepped up to the plate, and she was ready for me. When the show had begun I was only there with about five percent of my attention. The other ninety five was with my friends. So, she answered with exactly the same measure of attention.

Animals have hierarchies that are structured to ensure their survival. If there is a leader in the group, or a more dominant animal, the followers will allow that animal to maintain its status, as long as they are strong enough to assist with ensuring that follower's survival. The only way they can do that is to challenge that leadership from time to time, and particularly at times when the leadership is not as present as it needs to be. I experience this challenge often with the people I manage. When I am feeling at my worst, those are the times my staff will test my authority and rebel. All they need is to feel secure, to ensure that I am still at the helm. I experience the same questioning and instability in relationship with my children. When I am tired and have had a hard day, they are vicious in their attempts to get my attention. In the example of Frodo, she was doing the same thing; checking to see that I was still present in relationship with her, challenging my position in relationship with her.

Proactive relationship building

I have been taught in situations like the one with Frodo, that the way I choose to be in relationship with an animal will determine the outcome of training that I initiate with animals. Basically, what you give is what you get. This is one of the reasons I choose to use a reward system when working with the animals I train. Not only does this method provide a common language with the animals, but it also keeps me focussed on what is good. If I am focussed on what is going right, rather than what is going wrong, and having to acknowledge what is right every time I see it, then I am more present in the constructive side of the session. I am being proactive, rather than reactive. It is on this basis that I would like to define the type of relationship I wish to have with an animal. Furthermore if an animal or person is allowed to choose to participate, and then is rewarded for doing what is required, that same choice to comply will probably be made in the future. So, I am in relationship with an animal that chooses to be in relationship with me.

If on the other hand, no choice is made and an animal is simply coerced into submission, no real relationship to speak of is created. It will do that behaviour again, but only as long as the punisher is present. There is no thought complexity or choice in this. Very often it can be said that an animal is not even thinking about what it is not doing, only trying to avoid the punishment. This has been an extremely valuable lesson for me with my children. Because I have recognised that the animals are no different to me. My rebellious years at school, and the countless times I found myself sitting in the principal's office as a result of some trespass did nothing to change my behaviour. I just became better at not being caught doing whatever was considered inappropriate.

It is appropriate at this juncture to elaborate that in training lingo, the word reward is termed reinforcement. The word reinforcement means – to make stronger. This refers to the behaviour we are focussing on training. The opposite of reinforcement is punishment. Punishment makes a behaviour that we don't want to encourage weaker.

Basically, when I am present in the moment with my kids, I always have fun and they enjoy my company. There are no expectations of each other in those moments. And yet, in those moments I am not compelled to nag them to clean their rooms or help feed the dogs. We just go about getting the jobs done together. It was the same with Frodo and me in that show. If I was just telling her to do something so I could look good in front of my private audience, I was not in an unconditional relationship with her. Conditions were attached to her performance.

The lesson here for me is that giving and receiving are the same thing when we operate in the moment. There are no expectations or conditions attached to just being in relationship. On the other hand, if I am in relationship with anyone because I have something I wish to gain out of that relationship, the relationship will never be true. Because I will only give on condition, and probably receive, with the thought in mind – that I don't necessarily deserve what I was given.

I watched a child at the riding school who absolutely loved to ride. She had a gorgeous little pony and the two of them would go helter skelter around the practise jumping arena. At the end of these sessions this little girl would be flushed and

Chapter 1 Leadership – Giving and Receiving

happy. Pride did not enter into the equation. She just had fun. Till she started to compete. Her mother would stand at the side of the competition arena as her daughter and the little pony were preparing to go on, and nervously offer advice and direction. The competitions never went well. It was as though a different child and horse were out there. They were not having fun. They were trying to prove themselves.

If you are giving because you wish to give unconditionally, you are getting anyway. Any animal lover will tell you that. It is an honour and a gift to be in relationship with an animal. Gratitude and humility will bear more fruit than any successfully trained behaviour or rosette or boyfriend you will gain as credit to your name. The animals know this. Deep down we know it too. I have worked with a number of wonderful people in the past who have used their jobs with the dolphins as a "pick up line" to impress potential significant others. These trainers, wonderful people that they are, did not make careers out of working with animals.

Leaders with whom We wish to be Associated

For me, an effective leader is one who enjoys what they do, and who leads their people with that same attitude. If we take note of our own lives, we will be able to identify people for whom we enjoyed working. An English teacher who simply had a motivating exciting way about her, a mayor that made you want to contribute to your neighbourhood, a President like Nelson Mandela, that inspired you to be a part of a great nation. Then there are the authority figures who make us want to rebel; the ones whose manner is basically 'my way or the highway', dominators who instil fear as the impetus to action. When I have found myself in relationship with people like this, I am simply intent on finding a way to get my own way. Even subtle domination is dangerous. When responsibility for my actions is removed by an overzealous manager I do not willingly participate in the action required of me.

There is a wonderful story of a group of animal rehabilitators that demonstrates my mindset in this regard. This staff used to choose to come in after hours to help with animal rescues. Managers recognised this, and decided to compensate the staff with time off in lieu of the extra time spent at work. The result was that the staff no longer willingly attended the rescues. The reason? The staff, at some deep level, felt as though they were no longer the accountable personnel in the exercise. The managers had effectively taken away and assumed the employee's sense of duty to the animals.

An effective leader, on the other hand, is one who creates self confidence in their followers, one who generates participation. It is usually the leader who does not need leadership education – because they are doing what they do because they love to do it, not because they are trying to get it right. I will willingly follow someone like that, even if they mess up, I will forgive them because I am clear that their heart is in the right place. I easily worked for that English teacher. Homework in the subject was inspirational, but no different to the work I felt forced to do for other teachers. The manner of the inspirational teacher made me feel like what I had to contribute was worthwhile. Furthermore, the responsibility for doing the homework was somehow mine. It was not a case of do it or else. I had choice. As a leader we need to ask this question of our followers – 'Would I take you to war? Do I trust you to have my back?' If the answer to these questions is yes, then we are leading our team members. Basically, they will look out for us when the chips are down. In relationship with animals, will they still work with us if we are not coercing them into co-operating?

Many horse training methods propose a 'do it or else' methodology. The first one I was subjected to with my first riding instructor is a perfect example of that manner of training. Over

Chapter 1 Leadership – Giving and Receiving

centuries, horses have been subject to this type of training, with some wonderful exceptions. Traditionally equine equipment and technique have been used to force a horse to submit to pressure. The mistake many horse trainers have made, is forcing the issues, rather than communicating effectively with the horse. It is easy for horses to understand that they must move away from pressure. A foal can be taught to do this at an early age. It is what that foal would have to do in a herd if a more assertive horse came his way. For this reason, the existence of 'natural' horsemanship techniques is as old, if not older, than the methods that employ more aversive techniques.

The 'whisperer' methods are subtle, gentle, and are based solely on the premise that the horse will find it motivating to follow a good leader. I had heard about this technique of training, and visited and met a few individuals in my area who proposed the methods. What I did not get from the people I met was a method that made any sense. I saw demonstrations, but nobody managed to explain what they were doing in a manner that I could grasp.

It was not till I met Wayne Nicol that I began to understand how to use the horse's natural inclination to train it effectively, and as he says, gently. He was one of the teachers that Gandalph directed me toward. Wayne developed a company called Horse Gentlers International, and after learning from this master for almost a year, he invited me to join his company as one of his trainers. Wayne is an incredible mentor, and has taught me a great deal, personally and as a trainer. His humility, and eagerness to share all the knowledge he has is inspirational. That willingness to share information is because he is doing what he does for the sake of the animals. He cares deeply about ensuring that the unnecessary abuse that humans have caused horses is eradicated. I have seen him work many horses, some in really difficult scenarios, and always his approach is clear and effective. When working in this manner, the horses will remain attentive for periods of an hour and more, without any restraint. In the initial lesson that Wayne teaches the horses, he develops a communication system with them that has them, sans halter, follow him around, head lowered, eyes focussed, just like willing dogs. And that willingness continues throughout the relationship.

Wayne empowers his students with the same abilities. He abhors the term horse whisperer, and says that horses shout, they don't whisper. He demonstrates this when he teaches how the body language of horses is very direct. In my initial work with Wayne on one of his courses, I took Gandalph along, and was amazed that a whole new world opened up for me in my relationship with my little horse. At that point Gandalph was over two years old. We had a relationship where I would groom him in the stable. After one lesson with Wayne, I could look into Gandalph's eyes as he followed me around the round pen, and I noted something that had clearly been there all the time. Gandalph was looking to me for guidance. He was eager to be in relationship with me. He saw me as a leader and protector. I was awestruck and further humbled. There is nothing nicer than holding my baby's calm head in my hands, because he wanted to be there. I will never get over how incredible it is working with this level of attentiveness in horses in these sessions.

When I think back on my childhood lessons where we were constantly chasing horses to catch them, kicking them really hard to make them go and pulling really hard on their mouths to make them stop, it is no wonder that I developed a little fear around them. And the fear I developed was what my relationship with horses had been based on until I met Wayne. The reason why Gandalph pulled out of my hands and took off that fateful morning was his simple interpretation of what I wanted from him. Gandalph interpreted my intention with fear, and his natural flight instinct kicked in. Many horses who respond in this manner are then labelled difficult. It is only them giving back what they are getting from us.

Chapter 1 Leadership – Giving and Receiving

Using simple communication systems I see them now, watching intently, ears pricked up and eyes focussed directly, as they work out what we are requesting. The fact that the horses are eager to be with us, and able to remain focussed for lengthy periods is a clear indication that our leadership qualities are reinforcing in themselves. Every time I work with a horse, or see others utilising this method with horses, I am always a little emotional at how forgiving this breed of animal is. These powerful creatures can do great damage, but it is rare that they do. And when they do, it is generally that we have taken away their last bit of choice. In forgiving us our fear and clumsiness, they have allowed us to literally abuse them with the traditional horse breaking training. And there is an easier way. When I enter Gandalph's paddock now, he trots over to say hello. And I smile and am dazzled at my incompetence to hear what he was shouting before.

I remember watching Wayne saddle and back a three year old thoroughbred race horse in a couple of hours. Wayne had been called in to help the race horse trainers because they had been unable to get a saddle anywhere near the horse. In fact, they had gone as far as sedating it while it stood in its stable, and then placing a saddle on its back. That had only served to make the horse more afraid of the saddle. In a last act of desperation, the alternative being to have the horse sent to the slaughter house, and probably only because this horse was from noteworthy breeding stock, the race horse owners called for outside help. The horse did not even break into a sweat when Wayne worked with him. A simple act of communication was developed, and from there, the motivated attentive horse and consistent communicative trainer worked together to accomplish what is actually a simple task.

With the communication system in place, it was easy for the two to understand each other, and therefore, easy for them to be in relationship, working towards a goal. When Wayne eventually got onto the horses back, as is usually the case with the horses trained in this manner, the horse did not budge, and it did not have a halter or bridle on. It simply understood that this was the next step in a process that had been effectively communicated along the way.

What was sad, was that Wayne did this work with none of the race horse trainers watching. They were not interested in watching him, preferring to stick to their age old methods of training. However, they had acknowledged that he was the last ditch hope for this horse.

Pecking Order Misconceptions

For too long people have defined animal behaviour in a simple hierarchical fashion. One dominant animal over the herd. For many years, trainers tried to mimic this mistaken observation. They tried to be the overarching boss, demanding respect by forcefully and often cruelly keeping animals in their place. In my memory of a circus I visited when I was a child, one feisty lion tamer stood in his Tarzan outfit with a whip in his hands while the lions cowered on their individual podiums. Trainers of old used to teach bears to dance in the circus by putting them on a hot stove. The uncomfortable heat made the bears jump around.

Proper observations into animal social groups yield a very different and more complex picture. When we consider wild groups of animals, it is obvious that the group structure could not be effective if it was a simple linear design. There are different tasks that are required to be fulfilled in order to maintain the survival of the group; scouts, protectors, mothers to name a few. In a dolphin pod, groups are generally defined as matrilineal. The females are responsible for determining the speed at which the group moves as a result of their empathy with their offspring's needs. Their maternal instincts would also be a driving force when it comes to protecting their young. Surveyors on the outskirts of the pod determine where to feed and when to flee if danger presents itself. Dominant males fight for mating

Chapter 1 Leadership – Giving and Receiving

rights. But there is no clear sole king and leader. Some animals are more assertive than others, but generally they form a team of individuals working for survival of the group.

We did an observation of the group of penguins in our penguin colony at work just before we moved them to a new exhibit. We wanted to ensure that they all remained fit and healthy after the move, so wished to find subtle baseline behaviour trends that we could easily track. The birds are fed by hand, and every fish they are fed is recorded in a day sheet. This sheet has all their names typed out on one side, and spaces left for each feed so that an accurate account of their feeding behaviour and other behaviour is always recorded. We found that the birds seem to eat in a semblance of order. The same group of animals always eats first. The stragglers are always made up of the same individuals. We decided to use this baseline data as one of the measures to determine how the group was fitting into their new home. As it turned out that same feeding order was maintained. We used this information to help us determine that they were settling in well.

There were some changes however, because some of the apparently nicer nesting sites were further away from the feeding areas, and the more experienced breeding pairs had claimed these sights. They seemed comfortable feeding further down the pecking order as a compromise to achieve those nest sites. When we introduce new birds into the colony, we do observations. It is always interesting to note where they fit into the 'pecking order'. Males and females also respond differently. It is an easy assumption of many of us that most animals fit into a linear hierarchy, and observations such as the one outlined here are often what mistakenly generate this opinion. It is important however, to remember that animals' social groups cannot be this simplistic. There is not one dominant animal leading down the social ladder to the most submissive animal. Their lives consist of so much more, and when trying to understand this we need to be clear that the functions that determine their survival must be taken care of in the group.

Dominant breeding penguins, for example will not be the first to enter the water in most wild African Penguin colonies. In fact, the rule of thumb is to send in the juvenile birds first in case there are predators about. In the captive scenario feeding was on land, and the juveniles were not always afforded the same 'honour' as a result of the adjusted management strategy. The juveniles are also more concerned with investigating the area for the first couple of years of their lives. There is also a marked difference between females and males of breeding pairs. For instance, the males are much more territorial, and so protect the nest sites more vigilantly.

We looked after a group of farm birds at home for a while. Five ducks, two roosters and three geese. In general social interactions the geese were most certainly dominant. Then there was Hagar, the rooster. The second rooster was Blackie, an older male. He was dominant over Hagar, even though Hagar would interact more aggressively with us. However, when we would go to feed this mismatch group, the ducks always came in first to eat. The birds who were dominant in relationship with us would stay on the outskirts and wait for the ducks to finish. Horses too seem to have a clear pecking order, but if you watch a group of horses over time, you will notice that the animal that appears to be dominant, often named the alpha horse is not a very popular horse. He will stand on the outskirts of the group, and not have much social interaction with the other horses. If the horses get a fright, they will not follow this horse out of trouble. They will usually follow the confident plodder. The role of the alpha horse is not necessarily to lead the animals. Bonds between animals are also often interesting political associations. This has been well documented in primates, but exists in all species. Alignments often ensure privileges for the stronger animal and security for the other.

If we can assume the role of confidante, that confident plodder, the one that the animals look to for security, we have

Chapter 1 Leadership – Giving and Receiving

utilised their natural motivation to seek security, their survival instinct, and made it easier for them to trust us. We are no longer that erratic lion trainer with a chair in his one hand and a whip in the other. We are not the alpha horse, the dominator. We are the leaders, the consistent and dependable trainers.

Make no mistake, these social animals will also yield to that dominant alpha force, if we choose to assume that role. Recognise however that that alpha force is an erratic inconsistent one that employs fear as a motivation tool. If you scan the international animal training web sites, list serves and other news and periodicals that are used, you will note that many elephant trainers are killed by their charges, even in recent times. Is this any wonder when you consider that punishment forms an acceptable tool in many elephant training protocols? Eventually there is always the chance that the elephant will learn that those tools are avoidable.

In bottlenose dolphin social groups, it appears that there are groups within the school that make up the whole, and these groups all have different functions that work to ensure the survival of the whole. Recognising the intricacies of the social animals is important when determining an animal training programme. Furthermore, we need to determine whether they are flight or fight animals by nature, necessarily affected by the fact that they are either prey or predator, or both. These aspects and many other traits that are a result of the natural history of each species heavily influence the animal's predisposition in our training programmes. Generally, social animals have the natural capacity to yield to an authoritative force that is accepted as a leader. If we effectively assume the role of leader, the animals have the natural propensity to accept us as such. This fact is recognized and strongly guides most successful natural horsemanship practice. Experienced horse trainers will tell you that a horse will transfer his allegiance to an effective human leader because it is not unnatural for that animal to submit to an authoritative figure. This was my clear experience when I dejectedly watched the accomplished student work with a horse I had been trying to ride. The horse was clearly more comfortable under the student's leadership than he was under mine.

Assertive, Aggressive or Submissive

This scenario was a short while after one of my first horse riding lessons as an adult. For my first lesson, I arrived, apprehensive and excited in my brand new boots and a helmet that my husband had bought me for Christmas. Intent on rediscovering my love for horses with all the new insight I had achieved in my lessons as a marine mammal and companion animal trainer. I was introduced to a horse called Rum. A stout South African breed of horse called the boerperd. Rum is a slow moving horse, bordering on stubborn, which suited me fine. Any high flying forward moving animal at that stage of my confidence would have set me back. I felt confident in the basics of riding, but was about to learn that I knew very little. My first lesson was that when I mounted up confidently onto Rum's back, I had actually slumped into the saddle which was in effect very uncomfortable for the horse.

Dave, my instructor made no bones about telling me to literally lighten up. "You will hurt the horse if you just flop onto his back like that," he complained. I must admit that my pride was hurt, however I was impressed that my instructor was looking at the subject of horse riding from the horse's point of view. This was very novel in my experience. I had only ever been told where to pull and kick; the accelerator, steering wheel and brake, as a teacher I once had put it. The lesson continued along this vein. A couple of weeks later I had made some progress, but kept hearing Dave's words encouraging the horse forward. Rum is also a carriage horse, and I had seen Dave driving the horse in this scenario. He was using the same tone, words and encouragement to egg the horse on. I was

Chapter 1 Leadership – Giving and Receiving

beginning to get the impression that Rum and Dave were taking me for a ride. I was possibly not the one in control here, and furthermore, Rum appeared to be plodding more than trotting. I also found it very difficult to get him to stay in a canter, and when this happened, Dave's encouragement was pretty fiery. My concerns about my skills were soon to be illustrated.

One of the more experienced students arrived and was watching from the side of the paddock. I asked Dave a question about Rum's canter. I wished to know why I was having trouble maintaining the pace and why the horse kept slowing back into a trot. He asked the student to step in and demonstrate. I dismounted and handed Rum over and watched. I was amazed to see that Rum was behaving as though he was a completely different horse way before the student even asked for a canter. His head carriage changed, his walk was faster, he was responding to the slightest leg pressure, and seemed completely pert. His ears when I had been riding had occasionally flattened which is the mark of an annoyed horse. With the student, his ears seemed to be attentive, almost confident. The same horse, ridden by two different riders was performing very differently, even though we both appeared to be utilizing the same cues.

I have noted this often since that day. Generally, the horse will respond much better to the experienced rider. For sure an experienced animal trainer can potentially intuite what is happening in a training session and then guide that session using gut instincts. But while watching the session between the student and Rum, it seemed that Rum was able to intuit something about the student. The horse was responding to the person in a different way than he responded to me, and, for no apparent reason. I have also noted this with other animals; dolphins messing new and unconfident trainers around for no apparent reason, seals mugging unconfident trainers for food, even a rooster. Hagar, the red rooster that resided at our home, would intimidate me as I was afraid of him.

On more than one occasion he attacked me in a flighted charge. He never challenged my husband.

It has been a fear of mine and many of the trainers I work with that we are dominating the animals. Those visions of whip cracking "my way or the highway" type training being everything that goes against our desire to work with animals. We seek a partnership with the animals. Just like many of us seek partnerships with human beings. The opposite end of the scale is when animals are totally confused, without boundaries, and completely insecure as a result of a permissive relationship that they suffer in relation to their humans. There has to be a balance. But how to describe and explain this has often been difficult for me when working with trainers. It finally hit home for me when I determined the true meaning of the word assertive. The word belongs on a scale.

DOMINANCE......ASSERTIVENESS......SUBMISSION

When working with the police horse trainers, most of whom understand Afrikaans better than English, I asked one of them what the Afrikaans word was for Assertive. He told me it is 'selfstandig'. This, literally translates as SELF-STANDING. This makes great sense to me. Dominance or submission are both acts that commence out of a concern of the ego. Both are conditional states. Perhaps we are, for example, afraid, and so acting out to prove our worth or even more practically, acting to defend ourselves. Independence, or 'self-standing', for me, means acting in the moment, and without an agenda. Responding to the situation as it is, and doing it unconditionally, simply to be in a clear relationship with the animals. Because, if we operate without an agenda, we are operating with clarity."

Playing is being Present

As a child I used to train our pets at home. I had a dog called Seamus, an Irish Setter. I trained Seamus when I was only around eleven years old. I did not know anything about

Chapter 1 Leadership – Giving and Receiving

training, but managed to teach Seamus how to sit, lie down and even do some agility work using props. In retrospect I see that all that I managed to achieve in these sessions, because of a lack of theoretical knowledge on the subject, was a result of an intuitive relationship that Seamus and I shared. He was as eager as I was, and because of our close relationship, we managed to have some fun. We were communicating. Using operant conditioning is a dynamic process that combines the application of relevant theory, and an intuitive understanding of what is required at a particular moment. Some trainers have that gut instinct and are able to succeed in training with relatively little theoretical insight.

My two sons have repeated the feats that I achieved as a child with our dogs. Most kids do. They are not afraid to try. They play, and the dogs enjoy the game. Children feel their way through life. Probably the reason they are so much better on computers than most adults. They are not afraid of failing. They don't over-think the process. Most adults think their way through life. As adults, what often occurs is that all our conditioned misconceptions about ourselves get in our way. We forget how to play and our interactions with animals are about the goal, and not about the fun of being in relationship with the animal. This child-like gut instinct with animals is difficult to elaborate and define because of its subjective nature. Whatever allows our gut instincts to develop can be summed up when we watch children playing fearlessly.

Personal Belief Systems

One of our older male dolphins was out of sorts, and not wanting to eat. This is a terrible worry. Dolphins don't generally drink anything. They achieve all their required rehydration from the fish they eat. We also had to administer the dolphin with the medication that the vet had prescribed, and this had to be done with fish. But he was refusing the food. I communicated my worries to his primary trainer, and watched as she desperately tried to feed him. I had seen this scenario before, so asked her if I could take over. What I had seen when watching her is that she did not believe he was going to eat. It was written all over her face, in the way she walked and the way she sat in front of the animal. She was clearly very worried about the dolphin, and wanted to pet and stroke him to make him feel better.

I took a completely different approach. I consciously took on the posture and countenance of someone who believed the dolphin would eat, and went into the session to treat him as though he was absolutely fine. I asked him to do some of his trained behaviours that I knew he would find easy. He did not do them with great energy, but he did eat immediately I blew on the whistle telling him he had succeeded in his efforts. The medication went down and he was on his road to recovery. He had responded to me taking charge of the situation, something mothers do when they need to inspire their kids to believe that everything will be alright. Mothers take on the confident posture of someone who will protect their charges. They are not faking this posture. They will do anything it takes. It is their role.

My grandmother was an absolute lady. I remember her in Marilyn Monroe type dresses, her lipstick always perfectly applied and she always smelled of expensive perfume. When she visited us when I was growing up, her presence would always add a touch of class to our haphazard homestead as she elegantly infused our house with a sense of grace; until she was met by our kitten. My grandmother was not partial to cats. And it was almost as if the cat knew this. He would ambush her at every corner and if she sat down, he would zone in on her lap as his next spot of respite with kamikaze intent. Cats are renowned for this. The question is why? Some say it is because they find these people unthreatening. The bottom line is that the cats are reading something in the posture of those people that makes them respond in a particular manner. How we interpret it is simply that, our interpretation. It is also interesting to see how dogs treat people who are afraid of them completely differently to those who aren't.

What is happening inside each of us that is affecting our relationships with the animals around us? In the animal training world, this is a common debate. Why do trainers sometimes fail in training sessions for no apparent reason? The use of training technique is always the first issue that is discussed at these debates, but it is often the case that training theory does not answer the elusive question relating to the failures being discussed. My lessons from the animal kingdom, so far, tell me a different story. What I bring to the table in those training sessions is fundamental to the success of what I am attempting. In fact, my personal feelings will actually affect the way I implement the theory. More than this, it will also determine if I am able to read the animal and thereafter set it up to succeed. After all, being in the presence of animals, and particularly training them is communicating with them. How can I be clear with them if I am not in a clear space? It is like me trying to get my children to clean their bedrooms while they are upset about something that happened at school, and I have failed to notice that they are upset because I am stuck in my own drama – in the middle of an argument with my husband. Chances of my kids cooperating are very slight. I am not present with my request.

Listening

Confidence is essential, but it is the confidence of the soul, not of the ego. Confidence of the ego is what I would term arrogance; bravado, which, in effect, is the opposite of humility. It is me working with Frodo in a show to 'show off' to my friends. Confidence on the other hand is me being clear enough to admit when I don't know, and acknowledge with gratitude when I do. The confidence and humility that allows us to seek and ask for help when we need it. Gratitude, because we can never feel confident in our soul unless we are being humble. And the door to humility is gratitude. This is because we can never know everything. When we are in relationship with any other being on this planet, human or animal, we are always only half the story. In the role of teaching other beings, we need to develop a co-operative relationship. If I wish to know where I am at, I need to recognise, in the moment, if I am taking responsibility for the scenario, no matter what the scenario may be. This will keep me present, and that will present me with many more options.

Communication is only effectively occurring, if it is occurring in two directions. How we listen is just as important as how we speak. When working with animals, how we listen is even more important, because they cannot talk to us directly. Listening to them requires as much awareness as we can muster using all aspects of communication, including watching their posture, their health indicators, their responses and using our intuition. We need to consider how they talk to each other, and recognise that they will try and talk to us in the same manner. Different animals have different ways of communicating. We must be sensitive to this.

My husband agreed to look after a friend's three dogs. I felt uncomfortable the moment I arrived home, worrying about

the effect these three large animals would have on our five dogs and three cats. The cats were out of sorts, and this was evident as I walked in that first evening. The next morning they had soiled inside. I went to work, and on the second evening, was not met at the door by the cats as I usually was. That night while I was sleeping, I woke up with a warm

Chapter 1 Leadership – Giving and Receiving

wet sensation on my chest. Hermione is a street smart cat that I rescued. She was squatting and urinating on my chest. I woke with a start, but noted as I woke that the other two cats, Max and Philli were sitting at the end of the bed watching her. Hermione was very effectively communicating with me. The cats were not free to go outside as the new dogs were a danger to them. I got the message. They were provided with kitty litter until we could find alternative arrangements for the dogs. They will always communicate with us. It is whether we are listening that is always the question. Sometimes their communication is subtle – or rather our ability to interpret it is lacking. Sometimes, as in the case with Hermione, the communication is very clear. They are able to communicate, and if we are effective in creating a communication system with animals, we are guaranteed success as trainers. We are also guaranteed a more fulfilling relationship with the animals.

Leadership vs Dominance

It is vital, at this point that we discriminate between leadership and dominance. It is basically cooperation versus coercion. This distinction is important while reading through the remaining chapters of this book. The distinction is vital when we consider effective relationship. The distinction is however not always obvious. Very often it appears as if we are achieving cooperation, and theoretically we can even justify that this is what is occurring, however, our feelings in the session, and our subtle actions are actually coercion in action. To make the distinction absolutely clear, let's define the two concepts.

Dominance is a relationship where one assumes a position of control over another. The motivation for an animal to act is the fear or avoidance of a consequence. When we consider historical political figures such as Hitler and Idi Amin, we recognise dominance. Generally, a dominant figure will rule with fear being the motivating force that spurs us to action. The same is said about animals. Dominant trainers will coerce animals to do something where the animal fears a consequence, and so co-operates to prevent that consequence. This is total dominance. It must be recognised, however, that there are subtler forms of dominance in the training world. I watched a dolphin trainer asking a dolphin to perform a tail lob. This involves the dolphin slapping its tail on the surface of the water a number of times. The dolphin would do two lobs of its tail and then stop. The trainer wanted the dolphin to do four tail lobs. The trainer would happily have justified her position, saying she was using positive reinforcement and operating with clarity. When the dolphin returned, the trainer would animatedly turn her back on the dolphin and wait for a period of time before turning around and repeating the cue. This scenario repeated itself before me for around ten minutes. The dolphin eventually got it and lobbed its tail three times. At this point, ten minutes later, the dolphin, who was surprisingly still participating in the session, received its first bridge and reward. This in effect is dominance training. The trainer was effectively saying to the dolphin, 'you will do as I say or else'. The dolphin also did not understand the trainer's request. A tail lob is naturally a behaviour that dolphins would exhibit in frustration. It was clear to me at the end of that session that the dolphin slapped its tail more than twice out of pure frustration. This is not co-operation.

Co-operation is when we effectively communicate our request, and reward behaviour on the way to that result. Insisting, in this instant would not be necessary. This is because we are working with the premise that the animal is motivated to cooperate. Not as was the case with the trainer and the dolphin in this example, where it was clear that the premise this trainer held was that the dolphin was motivated to please her or get food as the treat – pretty egotistical and highly frustrating for the dolphin.

In terms of the relationship we generate with the animals when we train them using dominance tactics, the behaviour

Chapter 1 Leadership – Giving and Receiving

modification achieved, is a result of a controlling and compelling force a trainer would use to coerce an animal into co-operating as is the case in the above example. There is no choice in the matter, and no co-operation. I refer to this as power play. It is me yelling at my kids to do something 'because I said so.' When I hear these words wishing to escape out of my mouth while in altercation with them it is a reminder to sit up and take note of the fact that I have entered dangerous territory where mutual respect and therefore willing co-operation is no longer a part of the equation.

Leadership on the other hand implies something very different. It refers to a relationship where choice exists. It is where one party inspires willing co-operation from another party. With regard to the relationship between animal and trainer, behaviour modification is as a result of a motivated animal willingly choosing to co-operate.

I watched a horse trainer clicker training a horse. The trainer was fairly new to the methods, and the horse, also new to this manner of communication was reflecting the obvious excitement that the trainer was exuding. The horse was offering all sorts of responses without even being cued. His ears were up, listening out for the sound of that click. At times both horse and trainer would even forget about the food reward. It was true leadership in action.

The difference between leadership and dominance, in effect, can be very subtle. In fact, it is usually only in the attitude of the trainer that we see the difference. If as a trainer we are standing in front of an animal and insisting on a response which is not occurring, we need to analyse our feelings. If we are taking the situation personally, we are probably heading into the waters of dominance training. Our actions will probably reflect the attitude. If however, we remain in relationship with the animal, and see its choice, simply as that, its choice, we are maintaining our leadership persona, and more than likely, because we are not immersed in the frustration of wanting our way, we will act with an objective attitude, look at the situation rationally, and then do what needs doing to ensure that the training session remains a learning experience for the animal, rather than a session that is at the mercy of our dominating power play.

More than humility, in many cases, for me, this just requires a sense of humour. If we can laugh at ourselves getting into a power play, we will be reminded not to be reactive. I have seen animals that appear to have a sense of humour about the power play. Frodo is a master at this. She is a female dolphin, a mother and a grandmother, the matriarch and probably the wisest animal I have ever met. She is generally very willing and eager, if she is being trained by sensitive people. When a new male trainer began working with her, he was eager to prove his worth as a trainer. He was required to train her how to jump up and land on her stomach, a behaviour we call a tummy breach. She managed to get the behaviour in one session. The eager trainer called me over to show me the finished product. Frodo sat in front of him and refused to budge. She has done this numerous times. When a trainer is in the moment and having fun with her she complies. As soon as the trainer's motivation is to show off for others, she sees right through this ploy, and shows them up immediately. She is super cool.

Chapter 2

Developing Leadership qualities in Ourselves

Maintaining rigid autocratic control depletes the energy reserves meant for noting and acting in accordance with the guiding voices of universal truth. Humility is the key that unlocks our ability to listen. Humour is the key to humility.

The previous chapter outlines why we aspire to be leaders in our relationship with the animals we train. We cannot be airy fairy about this concept, and need to remain cognizant of the fact that our training relationship with the animals remain conditional. This is a concept I battle to describe to the trainers with which I have worked. They will disagree wholeheartedly, and tell me that they love the animals unconditionally. I cannot disagree. I have seen the most extraordinary feats of love and commitment by animal care staff in the years I have spent working with animals; dolphin trainers sleeping at a poolside when they are worried about an animal, seal trainers staying up all night to carefully note the behaviour of an animal they are concerned about, animal care takers and volunteers working forty hour shifts to rescue and begin the rehabilitation process with four hundred odd penguin chicks. A friend down the road slept in his mare's stable so that the animal could be medicated on the hour. The examples are endless. Taking care of animals is a vocation, and not a job. However, theoretically at least, our training relationship with animals is conditional. This is because any reward they receive in a training session is a result of an action the animals have performed. Generally, even if we are not consciously training them we are still in control of their resources, and thus automatically in charge.

The success of the session or relationship we share with the animal depends on whether the animal views us as a credible leader. To maintain ourselves as effective leaders requires the application of techniques that will engender co-operation from the animals we train. In time if we use our resources wisely, we will secure a trusting relationship with an animal due to the fact that they associate us with the positive aspects we have brought to sessions. It is also important to remember that animals probably seek out relationships that are self-reinforcing.

This is not only for people in active training relationships with animals. It is also true for people who keep pets. In terms of the animal's natural predisposition, trusting their conspecifics and particularly their leaders helps ensure their survival. They would naturally seek out the security of someone or some animal they can trust. To develop this type of relationship and ensure that this remains the case, we use all the methods in our operant conditioning toolbox that will inspire clarity, motivation and consistency. Look at all the books on good managers as leaders, and you will recognize the words in the last sentence.

Chapter 2 Developing Leadership qualities in Ourselves

Pitiful Confusion

Despite the best intentions of so many animal people that I know, those who truly care for their animals, many make the mistake of dominating them rather than leading them, through their inappropriate care. A classic example is when I ask trainers why they just rewarded an animal for a sub standard behaviour, and they trainer will mischievously retort "Ah shame, they really tried." Sure, they may have "tried", however, when we reward inconsistently, we are not being clear in our communication, and at some point, the animal is going to suffer confusion and potentially frustration as a result. Unclear communication is what dominators do. Leaders will always let us know, with clarity and transparency, where we stand. This results in us feeling secure around leaders. When I run animal training workshops I ask participants to train each other. One person assume the role of trainer and the second the role of "animal". Often I don't even let them brief the "animal person", because I want that person to sense the potential frustration that the animals suffer when we train in an unclear manner. This means, that when that "animal person" in the exercise comes into the room, they are unaware that they are even going to be trained.

Very often the "trainer" and other participants in the workshop will begin to make sympathetic noises and usually these inappropriate responses further confuse and bewilder the "animal person". As sympathetic human beings we are often too quick to pander and pity people and animals. We need to be extra careful that this does not result in us confusing our trainees. Pitying others means we don't believe they are capable of achieving. If we don't believe they are capable, this will reflect in how we try to teach them – we will be less believable. And the result (we will prove it to ourselves, because as human beings we so need to be right) is that the subject deserves our pity. As animal trainers, this is tantamount to 'loving the animal to death'.

Touching Animal Souls

Many dog trainers will relate problems that dog owners have with their pets that are a result of the dog being confused about the relationship it has with the people and other animals in the home. Most often, any unwanted aggression that dogs show to people in its family and to other dogs that are kept in the same household is due to this confusion. Dogs, being pack animals will naturally form hierarchical relationships with those around them, and this includes other dogs and people. If the dog is able to dominate its owners, this can result in all sorts of problems.

A classic problem that people make over and over again is how they manage the problem of two dogs in the same home when they get into a dog fight. Sasha, our adult female dog is a very dominant dog. She has some Doberman in her make up and so is genetically more inclined towards aggression which means we need to be extra careful in the way we manage her. She will maintain her alpha status by beating up on our other dogs if it appears they are threatening her status. We have a little dachshund called Captain Jack Sparrow who is very often in the wrong place at the wrong time, and has been attacked by Sasha on occasion. To manage this situation has been tricky.

The first response of my children and husband is usually to shout, which will serve to generate more excitement and potentially make the situation worse. The second response is for Jack to receive a great deal of sympathy and attention, and for Sasha to be chastised and excluded.

Sasha will be confused at the excitement of the situation, which she may read as encouragement. She will be further confused at the lack of attention or punishment that she receives. The sympathetic attention that Jack receives can compound her frustration at his apparent success to usurp her authority. All this serves to further intensify her attacks on the little dog. I explained this to my family, and we have made an effort to manage the situation better. If Sasha is demanding attention when the other dogs are around, we will

Chapter 2 Developing Leadership qualities in Ourselves

give it to her before the other dogs. We do however ensure that we remain dominant over Sasha by exercising various tricks. For example, she may never exit a door before us. We always go first. The dominant dog in a pack would go first.

If a fight does happen, we ensure that we manage the outcome sensitively. The fights have lessened in intensity and don't occur as often. They only occur now when we have visitors. Our understanding of this is that the pack hierarchy has become dynamic again. Also the little dog steals hearts and attention much faster than intimidating Sasha does. He does get more attention from visitors than she does.

A note on lap dogs, which essentially little Dachshunds are bred to be. If a dog puts his paw on another dog, or is able to put its whole body over another dog, this is an act of dominance over that other dog. A dog standing over another dog is a dominant dog. When dogs submit to the authority of another dog, they will cower under that other dog and break eye contact. If a dog places his foot on your foot or any other part of your body, he is asserting his dominance over you. He is also assuming responsibility for your safety in the same action. This is the reason that lapdogs are so clingy. They spend much of their time with all four paws dominating their people. They are very protective over their people and this is because of that unwritten contract that their people have accepted. The poor lapdogs have taken on the monumental task of dominating and protecting a sometimes very elusive pack member. They don't want their people to ever leave their line of sight, and if they do, the dogs become very anxious, because a person who is not visible cannot be protected. Often lapdogs are also responsible for being overprotective of their people, launching out and biting anyone who comes near. We need to ensure that we are not creating unnecessary stress in the animals that we keep. A simple understanding of their natural behaviour will provide insight that will enable us to be better care takers.

Maternal Empathy

When working with exotic animals, the same considerations are required. However, we are not always well informed of their natural behaviour and often have to feel our way around. Grace and humility will always serve us here – knowing and understanding that there is nothing more fulfilling and no greater honour than achieving a relationship with an animal. We were called to the rescue of a dolphin that had washed up on a beach close by. This type of call usually heralds a difficult period, and at worst, the death of the animal. Dolphins wash up for a reason. They are air breathers, and if they are ill, they may end in shallow water where they do not have to fight the waves to come up and breathe. As usual, it was with trepidation that we set out on this adventure. I was three months pregnant at the time with my first child. This in itself was transpiring as an overwhelming personal experience for me. I had always been, till that point in my life, fairly controlled in any emotional situations. I was completely unprepared for motherhood, and more than a little worried about the prospect. So, I was totally taken by surprise when we arrived at the rescue scene, and I experienced total maternal instincts taking over.

The dolphin was a very young calf who only weighed in at around sixteen kilograms. The diminutive character was holding his own bravely, but I felt his anxiety and instantly wished I knew how to make him feel better. After settling him in a paddling pool, we eventually made the journey back to the rehabilitation facility. He was in pretty good shape, despite the fact that he had obviously lost his mother. We spent the day at the beach where he washed up but there had been no sign of dolphins in the area. Before we had arrived at the scene lifeguards had attempted to swim him out to sea about four times, but he continued to beach back onto the shore, over some treacherous rocks. Shortly after he arrived at our facility, we named him Victory. The name was our intention to ensure his survival. He became known as Vic.

Chapter 2   Developing Leadership qualities in Ourselves

We soon had him suckling on a formula. We were concerned about his social well being, in that he had to be kept on his own till he was eating and for quarantine purposes. For this reason we spent a great deal of time in the water with him. Dolphin calves spend much of their time lying underneath their mothers. For this reason we also provided Vic with an inner tube from a tractor tire. He immediately found sanctuary in this blow up surrogate and spent time when we were drying our sea soaked skins, lying quietly underneath this tube. He became very affectionate, and if we were in the pool, he would swim close by. I thoroughly enjoyed the contact I had with this little calf.

At one point, after he settled, we reached a juncture in his progress. He had to learn to navigate from one pool to another. This task, if not learned willingly, can be taught by placing a net or barrier of sort behind the animal to literally frighten it into the next space. It must be understood, from a natural behaviour perspective, that it would not be wise for a dolphin to enter through a smaller entrance into an unknown area unless they were pretty confident about where they were going. We needed Vic to move from pool to pool, but did not want to frighten him into the next area. We could not reward him with food as he was still being tube fed his formula, and this procedure was not instantaneous enough to show him that he was doing the right thing at the right time.

I took on the task of teaching him this lesson. I made the decision before I began that this lesson would not be coercive. Teaching Vic how to negotiate this adventure was when I learned the power of relationship. I spent more time in the water, and I am sure I enjoyed this even more than my new friend. He would swim past me and initiate strokes from me, much to my delight. This involved him rubbing his flanks on my outstretched hand. He really seemed to enjoy the physical contact. This would be my currency. It was clear that he enjoyed this contact, and further to this, it was also clear that whenever I was in the water, he spent his time close by.

Gradually I moved closer and closer to the adjoining pool entrance. He did not miss a beat. It took a day to get him right up next to the gate. I stepped through the gateway, but continued to stroke him through the gate. I imagined he would just continue taking the strokes from my hand that was stretched through the gateway into his familiar area.

I had hoped this move would give him the opportunity to slow down at the gateway, letting him see that there was nothing untoward that would frighten him in the new area. I did not think that this little animal, without hesitating, would simply follow me into the area. This he did. No hesitation. He was visibly a little anxious, as his breath rate rose immediately. But, he trusted me, so came with me. It is important to note that this area had been open to him for a week before this session. He had access to it, but had not ventured into it alone. All it took was the power of a trusting relationship for him to make his move. I was blown away. To see this relationship was a wake up call for me. It is a fantastic responsibility to work with animals. I had, until that point always thought that the only currency I had was the food, or a secondary reinforcer like the tactile contact that this animal seemed to enjoy so much. I had no idea that the relationship we shared was that powerful. With that power comes so much responsibility. If an animal trusts me, it is my duty to ensure that I am consistently trustable.

With this added responsibility of being a leader in mind, let us discover some of the requirements of maintaining a trusting relationship with the animals we train.

Mean Yes When You Say Yes

A bridge is the way we tell the animal it has done the right thing. It can be something very technical such as a blast on a dog whistle, which is what we use with the dolphins. Many people use clickers with their animals. Clickers don't, however, have magical powers as has come to be the belief

Chapter 2 Developing Leadership qualities in Ourselves

in some circles. The use of the 'bridge' is what is important. It is there to be a clear part of communication, which is why the clicker can work really well. A simple 'good dog' can be a bridge. If we apply the bridge consistently and reward it appropriately it will become a powerful reward on its own. For instance, if a dolphin knows that a bridge is followed by a fish or a stroke, and a horse knows that compliance will result in the release of pressure by the rider, they will respond to the bridge favourably. This ensures that the communication system is always clear. The successful implementation of any type of training programme is in effect clear communication between the trainer and the animal. The animal knows what to expect and so trusts the process. In fact, I have seen some dangerous results when an animal is no longer listening to an effective bridge. It is very tempting to overuse the bridge.

When training an extremely obliging dolphin called Kelpie, the whistle bridge had been over used by the trainers. In effect, this is the trainer saying well done and 'yes' too often. It must be the same experience as when somebody over compliments you. Eventually you no longer consider them sincere in their compliments. They become bland and boring with their compliments. Kelpie was born in the facility, and the whistle had always been associated with good things for him. He was very conditioned to listen to and respond to it. His trainers adore this gentleman of a dolphin, and the overuse of the whistle had become an anthropomorphic attempt to tell him so. But the day came when he stopped listening to the whistle. In fact, it was clear from his responses that he was deciding when a behaviour should end. For instance, if we give the dolphin a signal to do a tail walk, the dolphin responds by moving backwards with most of his body out of the water, as though he is walking on water. Kelpie began ending the behaviour when he chose, by bowing out of the tail walk and rushing back to us.

This bow out of the behaviour was the movement he usually made when he heard the whistle. It became a real struggle to work with him. He was no longer listening to the trainers. A couple of the trainers thought he was deaf. There were investigations into hearing tests for the dolphins. This turned out to be an excuse. He was just badly trained.

The seriousness of the situation became evident when he excitedly terminated a calm body presentation – a behaviour where he was expected to lie still until the whistle went. As he ended this behaviour at will, he pulled his body around, and his powerful tail flukes knocked his trainer's head as he turned. She had not had time to get out of the way, which is what the trainers normally do before they blow the whistle. The trainer ended up in hospital for a couple of days with a serious case of concussion. To solve this problem, all trainers were required to become consistent and clear with their bridging technique. In fact, we even added a defining hand point to the bridge, to make it really obvious that the behaviour we were requesting was complete. The whistle tone for this dolphin also changed, so we could be absolutely sure we were retraining the stimulus. We found that the result was instantaneous. We had not even noticed that there was an element of frustration in his training.

The day we decided to be absolutely clear about the whistle, and ensure that Kelpie was rewarded every time we used it, his entire attitude changed. It was as though he had recovered from an illness. He was eager and excited about his training again. His eyes shone and his enthusiasm was infectious. His understanding of the bridge is the basis of our communication system with him. The communication system broke down and as a result it became difficult to achieve clear results from him. If he had been a more assertive or high energy animal we could have ended up in a situation where he became frustrated or worse, aggressive. When the whistle was not effective we were certainly no longer the leaders in relationship with him. We were inadvertently frustrating him.

Chapter 2 Developing Leadership qualities in Ourselves

Be Clear that the Gift is Deserving

There is an ongoing dilemma in the sales world. Should people work for commission or for a basic wage or for a bit of both? Should waitresses be tipped? What works better to motivate employees? Do we rely on their eagerness to do the job or do we need to incentivise the work in creative ways. In the world of animal training, the same debate exists. We need to ensure that motivation levels are continually maintained, and sometimes this takes a measure of creative license. Our goal however is achieved when we inspire motivation to be led by us, by promoting confidence in the animal as a result of setting it up for success. Affirmation in the shape of reinforcement in many forms will automatically result in a motivated animal. If we reward the animal too much for poor responses, this will result in a lower motivation over time. If we reward too little, the motivation levels will also dip. If we reward at inappropriate times, we will confuse the animal. If we reward inappropriately, we will cause confusion, or superstition.

These scenarios are all examples of communication that is not clear. The consistent approach we have when responding to the action of the animal will strengthen their trust in us as trainers and their own confidence in their ability to succeed, and thus, their willingness to continue to participate, interact, and ultimately succeed. Their need to succeed is often overlooked as a motivating factor. A colleague related a story to me about a killer whale that bit a trainer who was swimming with him during a water work show. When looking at why the whale may have bitten the trainer, the management noticed that the behaviour where the bite took place had been a problem for some time. They realized that in the weeks just prior to the bite, the whale had been rewarded for that particular behaviour less than 50% of the time. The animal was not succeeding, and there was a level of frustration that occurred that led to this attack.

Touching Animal Souls

These are obvious thoughts, often more obvious with hindsight. Our mistakes in these scenarios are not always obvious. We do not always achieve these ends in relationships with humans with whom we share a common language. With animals, we need to relate in a language that we invent. The association of an assured outcome when they cooperate will translate into a relationship where they know they can rely on us.

When teaching a horse called Spirit to move forward with a rider on its back, an experienced horse rider, while sitting on the horse's back, was initiating the forward movement by moving his body posture forward, and placing pressure on the horse's neck using his hands. As soon as the animal moved its body forward, the trainer would release the pressure. Releasing the pressure was the reinforcement for the correct behaviour. In time, because the communication process was so clear, Spirit was moving forward with a simple change in the rider's posture.

The opposite was the case when a nervous horse trainer was trying to make that initial contact, often referred to as a 'join up', with a horse called Jenna. The trainer understood that the horse needed to be rewarded for standing still while that trainer approached. To reward Jenna, the trainer took two steps backwards every time he got near Jenna if she stood still. It became clear, however, that as soon as the trainer tried to take that extra step closer, the horse would throw its head around and run off. The poor trainer became very disheartened, as this scenario played itself out about six times. Jenna had learned the game. The horse had been reinforced the same way and in the same pattern over and over. The situation was rectified after we advised the trainer not to be afraid to reward less on the approach in the first instance. Secondly, when the trainer advanced that next step, they were told to make a much smaller step, reward that quickly, and then advance again. The communication between animal and trainer was cleared up. The trainer learned that the horse is much more subtle and refined in its language than we give it credit.

Chapter 2 Developing Leadership qualities in Ourselves

A Good Leader is a Reward in Action

Pairing ourselves with reinforcement makes us a secondary reinforcer. In Pavlov's dog and bell experiment, we need to become the bell. Association is a powerful force. If this is done consistently and positively, we will become an inspiring trustworthy presence in the life of the animal. If, however we fail to maintain ourselves as an inspiring presence, the animals we are training will not be inspired to follow our lead. While watching a dog training class, I became aware of a lady who was intent on calling her dog away from a game he was having with a child at the bottom of the training ground. I had seen the lady enter with her Border collie pup, and saw how she spoke to him continuously. She obviously cared for the dog, and was constantly affirming him and asking him to do all sorts of things. He spent most of his time ignoring her. She had lost her value as a motivating force, and in fact, he had assumed the role of leader in their relationship. It was almost as though she was trying to please him in any way she could. She clearly had very little confidence in her abilities, and so had accepted her lot as an ineffective trainer. When she was calling him away from the child, she did not believe for one moment that he would come. In fact, five minutes later she had made her way over to the child and chastised the child for distracting her dog. I imagine that the dog found the child a whole lot more motivating than the lady.

Simple is Effective

Communicating clearly what we wish to achieve is vital. If the animal is not co-operating, it is not motivated to co-operate for some reason. Nine times out of ten, this is the fault of the trainer. If the animal is not responding the way we wish it to respond, the first question we need to ask ourselves in order to discern why, is 'how we are communicating?'

Touching Animal Souls

The fault can be very subtle, but it is there. This is why it often helps to have other trainers at the session so we can keep an open objective mind on what is occurring. There are some scenarios where the animal may not wish to participate for reasons of its own, but these are generally pretty evident; for instance, if the animal is hormonally motivated to be elsewhere doing other things, or if it is not feeling well. While watching a frustrated trainer teach a seal to jump up out of the water to touch a target, I noticed that the seal was becoming confused, and rather than swimming in to launch up at the target from the side, creating a wonderful arc jump that would have shown off his agility to the audience, he was jumping up straight at the target and landing on his body, creating a big splash.

A couple of days earlier the animal had been making good progress with the arc jump, and I wondered why it had changed to jump straight up at the target. I noted that he was touching the target which he had not achieved a few days before, but the goal of showing off its bow at the same time was lost. When questioning the trainer after the session, it became clear that the trainer was trying to train two things at the same time. The arc and the target touch. The trainer had become confused about how to achieve both objectives at the same time. Fortunately, rather than continue the session the trainer had chosen to end it and reconsider. Trying to train two things at once is very difficult. We need the animal to know exactly what we mean, and if we are rewarding it for two different responses, it will become confused. If we consider the power of variable schedules, that potential confusion becomes more evident. The trainer in this example had to restart the behaviour, guiding the seal in from one side using an extra target, and this guide was only taken out of commission to move to the next step once the seal knew that the end of the behaviour was touching the target.

Chapter 2 Developing Leadership qualities in Ourselves

Justice will be done, so be Fair

A dolphin we were working with was trained to put his rostrum on the ball of your foot while you were in the water with him, and push you around. This is a really good fun behaviour to do with a dolphin. We call it a foot push, because that is what he is doing. He is pushing you around. The dolphin provides the impetus, and you are able to determine the direction of movement by moving your shoulders and body like it is a rudder. That is to say, in the case of Kani, this was the intention of the particular behaviour. We began to have a problem with Kani while doing this behaviour, because he became intent on pushing the trainer towards the side, probably because this is where the majority of the rewards had been offered during these sessions. The dolphin has to actually be trained to allow you to determine the direction. If he wishes to, he is much more powerful than you are, and he will go where he wishes to go. This was the case in point with Kani, and the challenge we were facing was to tell him to let us steer. The problem had to be rectified. But, we made a fundamental mistake when trying to rectify this problem. Rather than take the behaviour back to basics, and train it out of show time, we kept it in show time, and chose only to reward when the animal did not choose to push us where he wanted to go. When we looked back in the records, we noticed that in the three weeks we had been working to fix this particular behaviour, we had only rewarded it around fifty percent of the time. Furthermore, three different trainers had been working on the behaviour, and probably all been rewarding differently. This behaviour had probably become a frustrating part of Kani's show repertoire.

No wonder that it resulted in a serious aggressive incident, where Kani eventually got so frustrated he bit me during the behaviour, and held me under the water a number of times, sometimes for close to a minute at a time. I was lucky to escape without serious injury. He held me in these aggressive motions for five minutes in total before I was able to escape. As trainers, we had not been communicating in a fair and clear manner with Kani. Furthermore, the balance of our reinforcement was not sufficient to ensure that the behaviour remained positive for the animal. It took us close to five years before we comfortably swam in the water next to Kani again. Our bonds of trust with him, in this regard were damaged, and we had to slowly get back to where we were. It is important to note, that in some cases, when aggression of this nature has been displayed, one will never be able to go back to the same situation again. We were in a really vulnerable situation being in the water with a dolphin that weighed close to four hundred kilograms.

The level of frustration that we caused in Kani was potentially life threatening to us trainers. But this aside, it is not fair to cause this much frustration to an animal, even if it is not going to result in a dangerous situation. Frustration is not conducive to the maintenance of a positive relationship with an animal. It is our responsibility to keep our communication clear so that we prevent this type of frustration from occurring in the first place. Never assume that all is well. We need to take the time to look at what is being communicated to the animal from the animal's point of view. We need to remain the responsible leader in the relationship. The photograph on the cover of this book is Kani five years after the aggressive incident.

The Role of Body Language and Posture

It is common knowledge that body language between people is a powerful contributor to the success of any relationship. Notice how the leaders you respect carry themselves. Generally they are confident and open. We respond to this because we feel comfortable with the fact that we are being led by someone who knows what they are doing, that they are not set to deceive us, or do things their way, just for the sake of being in

Chapter 2 Developing Leadership qualities in Ourselves

power. They are in a respectful relationship with us rather than simply demanding a pre-determined action or labour.

In recent years some horse and dog trainers have changed their training techniques to employ more positive reinforcement into their training protocols. These have been combined in unique formulations with the traditional body language strategies and have been employed very successfully. These formulations, and in actual fact, in all fields of training, the body posture of a trainer, who is unwavering, confident, assured and clear, results in a more productive session. We need to recognize these descriptions once again, as the same descriptions we would use to describe a good leader.

With human beings, the majority of what we say is analyzed in conjunction with the body posture, tone, mood, and manner in which it is said. Only around seven to ten percent of the message is in the words. In the animal kingdom, with common language being non-existent, another manner of communication is required. Each species fits into their ecosystem, some are predators, some are prey and some fit either category at different times. In view of this, when viewing creatures of another species, the animal will naturally try and work out where on the scale from threatened to threatening, respectively prey or predator, they fit.

It is commonly thought in the horse world that they see their riders as predators. Predators in the drive to kill are naturally tight and pent up. Predators are fight as opposed to flight animals. Let us consider how we may appear as a predator to a horse. If we were feeling anxious it would manifest, potentially, as us standing our ground and gripping in tightly. A prey animal's response is flight. They leave, and are exerting energy aerobically in order to respond in this manner. So, when working with horses, it is easy to see why they become anxious when an anxious rider is on their back. If we are anxiously sitting on their back, or even approaching them in a nervous manner, they see a predator stalking them or crouching, ready to pounce. It is not as cut and dried as prey and predator however. This need for having a clear and firm leader is not only a prey animal's instinct. Despite the fact that dogs are generally regarded as predators, they are also in touch with the fact that they could be prey animals.

The same can be said of dolphins and seals. Furthermore, these animals operate in hierarchies that physically challenge each other to achieve superior positions. This means they are aware of and sensitive to changes in their trainer's demeanour. They are used to submitting, but also have an urge to challenge authority. This, naturally speaking is absolutely necessary. The challenge has to occur because a leader who is not up to the challenge is not a suitable leader in which the follower can trust. And the trust is required as a pre-requisite for that animal's survival. These species of animal are very sensitive to hierarchy, and very often their behaviour is a result of their drive to be dominant or submissive.

Dog trainers commonly note that it is as though there is an energetic current that the dogs can pick up that travels down the lead to the collar. If the trainer is hesitant or nervous, the dog responds immediately to that feeling in the trainer. If we look at how we interface with leaders and even peers in our society, we may note that there are similarities in our primal make up that create the same patterns in relationship between ourselves and other human beings. A trainer asked me to come and watch her work with a seal she was having a problem training. The animal was aggressing erratically during the session. I arrived early to watch the tail end of a session she was having with one of the older female seals with which she has a strong relationship. I watched

Chapter 2 Developing Leadership qualities in Ourselves

the end of the session and then we moved onto the animal with whom she wanted assistance. I asked her to carry out the session as though I was not there. I was immediately struck by the fact that her voice was a few octaves higher with the second animal relative to the older female seal with whom she worked earlier. If we note how we talk to people of various perceived classes or hierarchies, we will see that our pitch is higher when we relate to people with whom we have more formal or fearful relationships. The trainer was practicing this same instinct with the seals. This alerted me to look at the trainer's posture, and it became clearer how the trainer was coming across as hesitant with the second animal. We were able to solve the problem by attending to these concerns. Her adjusting her attitude to the animal assisted her to adjust her posture, and the seal noted this clearer leader and responded favourably.

Thus it is reasonable to assume that animals rely heavily upon non-acoustic communication when dealing with animals of the same, but also of other species. Since Nature has equipped animals with a keen sense of reading the subtle, non-verbal body posture of other animals, in order to ensure survival, it is reasonable to accept that the body posture of the trainer plays a significant role in how the animal responds to that specific person. This has enormous implications, as we are communicating with them far more than we may be conscious of. We may think we are simply asking a bird to fly to a second trainer. However, the bird may be reading that we are not standing as confidently as we usually do, possibly as a result of the fact that we are working with an unconfident second trainer. And the bird may fly off into a tree. And we will probably blame the bird, or the second trainer. Body language and posture have the potential to enhance, or deplete our technique. Our communication is not limited to the cues which elicit and signal behaviour.

I found it remarkable watching colleagues of mine work two ten month old tiger cubs. The cubs were a part of an interaction programme. The trainers had trained them to allow the public to enter their enclosures and touch the tigers while having their photograph taken. I watched as a pretty confident looking young man entered the enclosure and took part in this memorable experience. I noted that the tigers had been intent on the game of tug of war before this person entered. As the man entered, he spoke in a loud voice, and strutted a little. My first thought when I saw him was that he was pretty confident around animals. As I watched a little more however, I saw that there were occasions when he was a little fidgety and edgy.

I had begun watching him more closely because the tigers had begun behaving differently. They seemed cautious, and one of them actually left the enclosure to go and sit in his safe space – an adjoining area where the tigers were allowed to retreat if they no longer wished to participate. As the man left the enclosure, the second tiger actually stalked him, and the trainers had to step in and distract him away from the man. I chatted to the trainers afterwards and asked them what they thought had occurred. We agreed that what had appeared as a confident man was actually a man putting on a show of bravado. We also agreed that we had only become aware of this performance of fake bravery after we noted that the tigers were behaving differently towards the man. The tigers had picked up on the man's trepidation, and noted it way before any of us did.

An experienced animal trainer should be able to observe the body posture of the animals they train, and respond accordingly. For example, if one was about to offer a dolphin a cue to ask it to jump and the dolphin is not as alert as it usually is, with a head motion that is more to the side than upright, one could refrain from offering the signal. Set the animal up to succeed as it were by asking for an alternate behaviour that the trainer knows the dolphin will do, wait for it to look alert, and then ask for the jump behaviour. I have seen good dog trainers do the same thing. In agility sessions, I saw a South African champion get his border collie who was looking a little lethargic, excited and motivated with easy little behaviours before he set the award winning

Chapter 2 Developing Leadership qualities in Ourselves

dog on the track to do his showing off. Observations and research offer countless examples of the use of body language in the species under review. Watching and learning the body language and posture of the animals that we train aids in the success of their training.

Confidence in a trainer is no doubt automatically reinforced in the trainer, the more successful that trainer is. People who are used to working with animals usually have more confidence than others. And they will present themselves to the animal with more confidence, as a leader. Confidence affects the practicalities of operant conditioning because it translates into body language. To harness the power of confidence, it is necessary to see how trainer confidence can manifest in our training sessions and how it will affect the applied theory.

We Get What we Give

Inappropriate requests for an animal to co-operate are usually the requests of an unconfident trainer, or a trainer who is not confident that the animal will respond. The way we ask will determine what we get. I have seen and been the perpetrator of this crime on many occasions. When an animal is in front of me and I know in my gut they are not going to take the cue, but I present the cue anyway. And they don't take it. Because in the giving of the cue, I did it with less energy and no belief they would do it. If I feel they are not going to take the cue, I must not offer the request.

Sometimes belief does not even come in to the equation. Sometimes I am just low in energy, and the bottom line is – what I give is what I get. This is such a wonderful phrase, and can be applied to any training situation that you ever find yourself in. We conduct shows with the dolphins and seals, and there have been times when I have been tired while doing those shows. The effort that I receive from the animals

Touching Animal Souls

on those occasions where I am not fully eager and present in the session is always a reflection of the amount of energy that I am putting into the session. When training dogs, I have exactly the same experience. And if I am meek and feeble when riding horses, the horse responds with meek and feeble energy. The examples are endless.

I was working with a Sea Lion trainer at the Giza Zoo in Cairo. The Sea Lion, Lembi, was chosen as a training subject as he had presented as the most difficult animal to train. The Sea Lion, up till that point, had been baited into various positions in order to get him to comply. The problem with the baiting is that the animal had never learned anything except how to creatively get to his fish. His fishing behaviour had become so bad that when the trainer tried to feed him by hand the animal would snatch the fish from him to the point where he would graze the trainer's fingers. We decided to target train Lembi so that the animal would begin to learn how to think. I was trying to explain to the trainer what to do, and this was challenging as we were working through an interpreter, as the trainer, Mohammed, only understood Arabic. It was clear to me that Mohammed did not believe that what we were doing was going to work. He was hesitant, and probably a little unclear on what was going on, and this confusion was relating through to Lembi, who was behaving unenthusiastically. Eventually I took hold of the target and used the Arabic bridge – Gadda, and then Mohammed rewarded the animal by tossing a fish to him. Mohammed became clearer on what was required, and as I saw his eyes light up I gave him the target and stepped back. Lembi had no problem achieving the target training. The excitement on the Sea Lion and the trainer's face was priceless. In fact, as Mohammed stood up and had an exuberant conversation with the interpreter, he casually held the target at his side. I laughed when I saw Lembi patiently sit at the target for a good thirty seconds. I had to alert Mohammed to Lembi's compliance. He was even more excited when he noticed the power of the lesson he had just taught.

Chapter 2 Developing Leadership qualities in Ourselves

Grey is Confusing, Black or White is clear

In a beginner dog training class, a well dressed professional young man was calling to his dog to return after the exuberant spaniel had gone AWOL during the class. The animal had been distracted by a couple of hadedahs that had landed at the bottom of the training field. The young man looked the type who did not have anyone cross him. The dog came bounding back to him, as he had been asked, only to receive a thorough chastising once it reached its owner. How confusing for the poor dog. The man was shouting at the dog for being distracted, a behaviour that had been displayed about two minutes before. The dog responding to that all important recall was basically punished for listening to the owner. This young man had clearly not done this before. The dog was motivated by the man and came quickly when he called. But the young man being in a class, with mostly females, felt a personal affront when his dog took off, and took out his embarrassment on the dog. His lack of confidence was a result of his own insecurities. His erratic responses in training sessions, if they continued in this vein, would result in a confused and frustrated dog. This would not generate a desire in the dog to be cooperative.

Little Tricks of the Trade

Usually, in most successful training programmes, whether of domestic animals, or groups of exotic animals in animal facilities, there are either situations where each animal has a primary trainer, or where only one trainer trains at one behaviour in an effort to avoid confusion for the animal. Unlike with pet animals, it is however important that the animal learns to work with different people. It is not acceptable, in the work place, to put all your eggs in one basket. If something happened to that one person with whom the animal is in a close relationship, it needs to be assured that the animal will maintain his well-being and not suffer without that one trainer. So, in most facilities, new trainers are introduced to trained animals. I remember one trainer who had an excellent relationship with a seal called Sasha. (The seal that my dog Sasha is named after) Sasha and her were excellent together, and it seemed that the seal would do anything for the trainer.

When new trainers were introduced to the scenario, they would have very poor success rates. The seal just would not respond the same way. The new trainers were being taught by the successful trainer, but still, the problems continued. It took a senior trainer at the time to notice that Sasha's trainer was delivering the dry training. For example, she would say do this cue to illicit this response. What the trainer was failing to deliver, were the little things she knew about Sasha that served to make the sessions reinforcing for the seal. For example, Sasha was a high energy animal and responded better when worked at a fast pace. She responded very well to certain behaviours, and these, used in the correct instances, would aid in the success of the sessions, by seemingly maintaining her enthusiasm and focus.

Also, the body posture of the trainer was novel in that she always stood next to the seal, whereas she was teaching newer trainers to stand in front of the seal. The level of success that the newer trainers were experiencing only improved after they fine tuned their approach to mimic the primary trainer. We learn as much by watching people who already train, well and badly, as we do by training ourselves. In the case of learning one particular animal's repertoire, the more experienced trainer is already established as a leader, their body language and mannerisms in the relationship have been reinforced. Mimicking these small reinforced nuances, will help us to assume the role of leader. The little nuances are not the only things we can mimic.

When I have watched master trainers train, I often find that it is not what they say that makes the largest impact on me.

Chapter 2 Developing Leadership qualities in Ourselves

Watching what they do is often a larger lesson. For example, when I watched Wayne demonstrate his horse gentling techniques, he never once mentioned that it was important to reward the horse with a stroke. However, he always does this. In the formative stages of training green horses, he pairs the stroking with a release of pressure. For example, when he asks a horse to yield its head to the side by bending its neck, as he lets the horses head go he will gently stroke its neck. This stroke becomes a part of Wayne's toolbox with every horse he trains. I mentioned it to him one day, and he laughed and said he had not even been conscious of the fact that he was doing this. It is now a vital part of my toolbox too. A simple way to tell the horse he has done the right thing. In effect the stroke is a trained secondary reinforcer.

Feel It

This subject will be elaborated on a great deal more under different subject headings during this book. The most important aspect to remember at this point, is that when we are confident and present in a relationship, we are able to feel into the situation, and literally be intuitive. If we are fearful, or more concerned about getting the session right, rather than feeling into the animal's state of being, we will not always be successful when it comes to setting the animal up for success. This is often the case with trainers who are still building relationships with animals. A good leader directs confidently assuming that their followers will succeed. This attitude of confidence inspires confidence in those followers.

Quality of your Attention

Managers that attend leadership training will recognize this one. 'When attending to a staff member in your office, ignore the phone when it rings. Make sure that your staff member feels valued, and offer them one hundred percent of your attention when necessary.'

Touching Animal Souls

In relationship with animals a hesitant trainer often looks at their peers or others in the vicinity, rather than maintaining appropriate visual and body contact with the animal they are training. A follower needs to feel that their contribution is important. Without this fine level of reinforcement, the other reinforcement received can lose some of its significance. For instance, a fish offered to a dolphin with encouraging body language is potentially more reinforcing than one offered without it. I was asked to watch one of the newer trainers work with one of the more seasoned dolphins, because they were having problems with the dolphin not wanting to cooperate. He was swimming off refusing to take cues, which was very out of character, as he is usually a stalwart performer. What I saw was the trainer looking for assurance from more seasoned trainers, and not fully focused on the dolphin. She seemed more eager to please the trainers than be in relationship with the dolphin. This was particularly the case when she delivered cues. She did so looking at the trainers, not at the dolphin. As if to say 'Why won't he do this? Is my cue correct?' Clearly it was not. She was not delivering it to the dolphin as a confident leader would.

The trainer was most distressed when this was pointed out to her. The dolphin had managed to point out an insecurity in her that ran really deep in her psyche. She was unconfident, and very worried about what certain people in the dolphin training team thought of her. She found herself in a situation where she had to choose whether she wanted to succeed as a trainer in spite of her concerns. She chose to succeed. In the very next session we saw her take the bull by the horns and offer the dolphin her undivided attention. The result was a co-operative relationship where the dolphin eagerly went through his paces. Fortunately the trainer noted this, and was able to laugh at herself after the fact. A sense of humour is a sign of a good trainer. This trainer remembers this moment as a transformational one, not only with regard to her career with animals, but also because it highlighted an aspect of herself that had significance in her life in general.

Chapter 2 Developing Leadership qualities in Ourselves

Taking Responsibility for our Actions

Trainers who lack confidence will unconsciously allow the animal to take the lead. This can cloud clear communication, and very often it can be very unsafe. If we are unable to control a horse's quick race back to home ground after a ride on the trail, we are not in control of the horse, and will be unable to handle the situation should a problem occur. A trainer who allows the fast exodus back home is letting the horse take charge which leads to difficulty in controlling the behaviour. I have seen dolphin and seal trainers make similar mistakes by letting the dolphins and seals crowd their space because they are eager to be fed.

The intimidated trainer feeds the animal to keep them from entering closer into that space, in essence rewarding them for this invasion of space. This can lead to assertive behaviour from the animal and even aggression. This aggression is a result of the boundary of leader and follower being crossed which results in confusion on the part of the animal, which leads to frustration. If you are the trainer, you are in charge, and it is up to you to draw clear boundaries, and maintain those boundaries with clear communication in order to maintain your status as an effective leader.

We worked some Appaloosa horses on a stud farm. The woman who breeds the horses has an excellent method of imprinting her foals. The mares are all calm animals. The woman has a great relationship with them. When the foals are born, she spends time with the youngsters from the first days. The result is that the two and three year olds that are being started under saddle, are confident horses. Because they are not afraid of people, it was more of a challenge for us to keep them out of our space. They were eliciting the stroking, which they obviously found rewarding. We had to be very careful to only offer this reinforcement when the horses were co-operating. One of the stallions saw us as his new favourite game. I watched Wayne struggle with him for a while before we all saw what was happening. The horse did not understand pressure and release. He only saw that everything was for his entertainment. He was not a very socialized horse, as he was a large and dominant stallion, and so we had to spend much more time teaching him the rules of our new game than we normally have to with other horses. He was not being mean. He was just looking for new fun and games. In fact, on two occasions I left my carrot stick next to his enclosure between his sessions. On both occasions he managed to stretch his neck out, grab the stick, and when we returned, we found that he had a new toy. He even played with the halter rope as though it were a toy.

Play Play Play

The stallion has a lesson for us. If we are having fun, everything else falls into place. This is one of the most important rules of animal training and another concept we will revisit often in this book. If we are not having fun, we are in the relationship for the wrong reasons. This has very often been the primary difference, in my experience, between successful and unsuccessful trainers. The animals pick up on the mood. An unconfident trainer will not naturally remain enthusiastic, clear and confident in the session and the variety and reinforcement value of the relationship between that animal and trainer, and the reinforcement value of the session will thus break down. The examples of this are endless. And it makes such sense to me. I would much rather be in relationship with someone who is fun. People who are serious all the time, and fixated on following their own pre-determined manner of doing things are not motivating to be around. They offer me no room for expressing myself. So often I get called to take a look at a problem behaviour, and what I see is a trainer who has become so frustrated with a session that they are completely in their head about it, and totally fixated on what they want the animal to do, rather than focusing on the situation as it is presenting itself right in front of them.

Chapter 2 Developing Leadership qualities in Ourselves

I watched a lovely young man working his horse, Ashaan, trying to teach her to move her hind quarters away from him in response to his body posture. Ashaan was clearly confused, and the young man's frustration at not being able to make progress was evident. So, the horse was doing what a horse does when it is faced with a threatening atmosphere. It was avoiding the trainer. The young man could not see that Ashaan was confused. He only saw what he wanted the horse to do to listen to him. I asked the trainer to leave the session. He did and went off to light a cigarette as he exited the round pen. We sat and chatted about what was happening, and then the conversation moved onto the evening before, where the trainer, a new dad, had contended with the baby crying with colic for much of the evening. He was tired. Fortunately his sense of humour presided, and soon he was laughing at himself and his situation. When he entered the round pen again, his softened countenance was obvious, and Ashaan responded within moments. He laughed long and hard at this, and I think that lesson will stay with the trainer for a long time. He recognized that he was feeling victimized when he entered the round pen the first time. This feeling was being expressed in his hardened frame. He was presenting as a victim. Ashaan was simply giving him what he expected. The world out to get him. Times when I have found myself in the same situations have been the most humbling, and in retrospect, entertaining. There is nothing more liberating than being able to laugh at ourselves.

Chapter 3

Sensorial Animal Oneness

The relativity of life is perceptible through our senses. The experience we live through our senses we call our own and yet is enlivened by the world around us; the world that is a part of us, and of which we are a part.

I have watched flocks of pigeons on the beach front in Durban while my kids have spent time becoming perches and feed bowls for them. It amazes me that the birds will hop about and one will fly off and return. This group of birds is pretty used to human folk and other distractions that would normally frighten most other winged species. What always amazes me about these birds is that one will often fly off and perch elsewhere with no great distraction to the other pigeons. However, occasionally one will take flight and this will spark the rest of the birds to follow suit. They are so in tune with each other, that fine posture can determine action. The dolphins too are so finely tuned to each other. The males have pair bonding moments where they synchronise their movements in far greater detail than any trained professional synchronized swimmer or dancer can ever master. Slight movements of their pectoral fins, a little nod of their head sideways, blow out a bubble and so much more makes up

Chapter 3 Sensorial Animal Oneness

their repertoire of behaviours, and seemingly it occurs at the exact same time. Such sensitive creatures. Dogs have been trained to sniff out cancer cells on human beings. Their sense of smell is way beyond ours. Some marine animals including dolphins are believed to navigate using magnetic fields. Much has been researched on a bats use of sound to navigate. Chickens are believed to see colour much more intensely than we do. Animals sensory capacity is different to ours, which means that they experience the world differently than we do. Animals have incredible senses.

I remember working with two dolphins and having a wonderful session practicing behaviours they already knew as a form of exercising them. Midway through the session, I was approached by someone with whom I was in conflict at the time. Immediately my session with the animals began to turn sour. One of the dolphins drifted off to chat to some other dolphins through a gateway, and the other sat stubbornly in front of me refusing cues. I have not met a trainer who disagrees when I say that a mood that we are in has the potential to affect the training session we are conducting. In our relationships with the animals that we train, it is proven that our state of being has the potential to affect the animal with which we are in relationship. How they read us is not fully understood, yet to remain clear in our communication, we need to be conscious of the picture we are presenting them. Our feelings and confidence and moods in the session are our responsibility, as they will affect how we read and deliver the session and thus, the way they interpret our presence.

The consciousness we have of our body language is a fantastic tool. Our awareness of the fact that body language and posture are part of the communication process in animal training sessions will make us better communicators in relationship with them. As animal trainers, this will enhance our ability as trainers. Being aware of our posture when relating to our pets at home will ensure that they are not confused and frustrated, because we will be clear. Many people practice these skills without even knowing that they do it. My brother in law came to spend some time with me at my home. He rode horses for most of his childhood, and competed at pretty high levels in various disciplines. His mother always said that he had a way with horses. He had been taught to ride in the traditional English manner, and so was accustomed to using bits. While visiting he rode our horses. He was totally amazed at the fact that we did not use bits on our horses, and even more amazed at how responsive the horses were. I explained how we had trained the horses all their cues on the ground. For example, we taught them to disengage their hind quarters, which means to move their back legs only, by initially moving their heads up in the direction of their withers, and then placing our hands in the place were we would eventually place our heel, just behind the girth. Eventually this would, on the saddle become a gentle pull backwards on the halter and placing our heel gently on their sides. The horses perform this exercise perfectly. They will even side pass if this cue is performed at the same time as a direct rein cue. What amazed my brother in law more than anything is that he already did a lot of what was required when communicating with the horses. He had however never been taught to do this form of communicating. In his years of experience he had worked out that what we were doing with the horses is what works. What he found, was that now that he was more aware and conscious of what he did that worked, he could more easily use the tools he knew about to achieve success. His sensitivity increased, and his ability to communicate with the horses grew. Happy horse and happy rider. In general, there are many more benefits this type of consciousness will offer. Let's investigate some of them.

Awareness Alone

When we believe that something is possible, the message we are giving off in our body language is exactly that. We are not in the midst of judging the possibility and we remain present, focused on the task at hand. Hesitating is clearly read in posture. We can read it in each other. Animals

Chapter 3 Sensorial Animal Oneness

who are necessarily primed to read posture 'for a living' – to ensure their survival will read even the most subtle signs. Posture and nothing else can be the difference between success and failure in a training session. So, even if we are unconfident, but sorely require a result, we can mimic this confidence – if we have generated an understanding of how it appears as a result of previous successes. We moved a group of seals to new premises. To achieve a stress free move, we had taught the seals to walk into crates and be carried around. We had practiced the move over and over and added in all sorts of variables that we thought we might need to be used on the day. We even used a truck engine running. Because of all this practicing, we had the experience of succeeding. On the day we had to erect scaffolding. That was unforeseen in our planning for the move. The seals had never seen the scaffolding. They sensed something different, and appeared hesitant at the start of the session. We noted this. We regrouped before beginning the process and discussed our concerns. We chose to conduct the sessions in the exact same manner we had in all the rehearsals. The rehearsals had worked over and over. We had experience of this success. So, we concentrated on succeeding. And this memory of success translated in our posture. We were not coming across as concerned and hesitant. In spite of the alien metal, we believed we could succeed. The procedure went off without a hitch. All successful trainers will tell you a time when using posture, they bluffed belief in an outcome, perhaps unconsciously, and achieved success as a result.

Training Trainers to be Confident

I will only go as far as I am comfortable. When grooming a horse, for example, unless I know the horse very well, I will not groom without a lead and a halter. This is to prevent a scenario from going sour if the possibility exists at all. It makes me more confident in the horse's presence, which in turn will allow the horse to feel more at ease than if it were to have a nervous person agitatedly moving around it. It is important that we don't let fear stand in our way of being confident. The animals read that fear in a variety of different ways, and for some of them, they will perceive our taut energy as a threat to which they may be forced to respond. Success inspires confidence, which will result in better trainers in the long term.

Managing the Dynamic of a Group

At times when working with animals, we need to work with groups of animals at a time. There are many different philosophies on how best to do this, but the bottom line is that it is vital that we maintain motivation and social harmony in that group. With these objectives in mind, we need to maintain our role as leader and take care that our body language does not convey a message that we don't intend. When we work with the dolphins, our attention is a secondary reinforcer. This attention is in the shape of our presence, and our eye contact. When working with two dolphins, we need to share this eye contact between two responsive animals.

A trainer I was watching was working two bottlenose dolphins, Affrika and Zulu together as a team. The trainer was Zulu's primary trainer, and unconsciously, in the session, with very subtle posturing, was favouring Zulu. This resulted in aggression between the two dolphins and a break down in their team behaviour. Affrika would drift off whenever this trainer worked the pair, and Zulu would become more aggressive in these same sessions. The trainer had to note her alignment with Zulu, and then work backwards to maintain harmony. She subtly began offering Zulu more attention whenever Affrika was in close proximity.

Whenever the pair were together they were eagerly reinforced by the trainer as she animatedly told them how wonderful they were. They would then be sent off on a high energy behaviour, something both animals find highly rewarding. On their return, because they had done something together, they would once again be met by a trainer praising them with excited posture and fish reward. If ever Zulu pushed into Affrika's space, the trainer would respond in a manner that did not reward the dominant dolphin. Either remain completely still for a few moments, leave the area completely, remove eye contact from both dolphins, stare at the space where she wished Zulu to station in front of her, or hold out her two hands onto which the dolphins are taught to station. All these responses are posture related. The trainer also had to be very aware of the subtle changes in the dolphins that signified that social dissension was imminent. She wished to catch the dolphins doing the correct thing and so did not wish for the aggression to occur in the session, rather wishing to prevent it from happening. This manner of managing this duo was extremely successful.

Trust in the Relationship

Consistency in our training methods, in all aspects of the communication we have with the animals will result in an animal that trusts us. If we inspire confidence in the animals we train, we will have greater measures of success with our training in general. The consistency is importantly, also about our posture.

I have seen many examples of this. In one scenario, years ago, a dolphin called Jula, with whom I am fortunate to be in a close relationship injured itself and had to be put on the stretcher in a hurry so that the vet could attend to the wound. There was a fair amount of stress involved in the procedure, and all the animal care staff were rallying around trying to get the dolphin to move easily onto the stretcher. He was resisting, and it was obvious that the scenario was making him anxious. Because I was pregnant at the time I was not involved in this potentially dangerous process. Because of the urgency of the situation, and because the dolphin was not complying, I eventually entered the water to help with the procedure. I placed the dolphin's snout into my hand and looked him in the eye. He visibly relaxed and the procedure went by smoothly from this point onwards. We were both confident in each other's presence. It must have shown in my posture, and was reflected in Jula's actions. This was a lesson that we used when we moved ten dolphins to a new facility. We ensured that all the accompanying trainers were people with whom the dolphins had close trusting relationships.

Body language is a powerful communication tool, and it is vital that we fine tune it at an interspecies level to ensure that we do it effectively. As leaders in this relationship, we are the caretakers with the ultimate responsibility for the well-being of the animals in our care. The inherent benefits of the stimulation and activity that we offer the animals during training sessions needs to be as clear as possible, to ensure that it remains motivating and reinforcing. Many of us who work with animals will, when we take careful note, realize that we have already been affected in our posturing. We know what works when asking animals for certain responses, and we duplicate the posture over and over. When we become conscious of what works, our jobs just become so much easier. And the animals become so much more receptive, because we are communicating with them effectively. Clear communication results in calmer, confident and at the end of the day even healthier animals. It is our duty, because, their well-being is why we do what we do.

Chapter 4

Why we do What we Do – Expressing Who We Really Are

As children we fearlessly display our true selves. If we quell that passion our trueness is packaged into unlived dreams. Until we awaken to the fact that our dreams are our talents we keep those dreams asleep. Caring about ourselves to ensure that we have the courage to live out our dreams is caring to make a difference on this planet.

I am so grateful to be an animal trainer. As a child I was enthralled at the possibility of talking to animals. I have watched the work of scientists who work with language acquisition skills in animals with total excitement. For me it is as exciting as travelling into outer space. There is so much mystery there. When I look into the eyes of an animal I am overwhelmed with questions. I want to know how they are feeling and what they are thinking and how I can do better for them. I believe that through the use of enlightened animal training techniques we achieve a level of communication with the animals that is probably beyond the honesty that we have in most of our relationships with people. It excites me and humbles me over and over again.

First Lesson

Even though I have trained a number of dolphin calves, the excitement of watching them learn our way of communicating never ceases to amaze me. I clearly remember Ingelosi (Zulu word for angel). At just around four months of age, he had recently learned how to stop, and he was beginning to solicit attention from the training staff. I set myself the task of training him his first behaviour ever. I needed him to exhale out of his blowhole on cue. This behaviour is trained so that we can send blowhole exudates samples for medical analysis so we can keep a close eye on their health and monitor their health proactively. He knew that the whistle meant good things because he lived with his mother, and she would return to the trainer for food and attention when she heard us acknowledge her behaviour with that stimulus. He was not yet eating fish, but would mimic his mother's excitement. He was an eager participant, and seemed to want that whistle to blow for him. When he heard the whistle, Ingelosi initiated stroking sessions with us, seeming to enjoy the tactile contact. For the session I simply sat with him stationing his little mouth on my hand and waited for him to exhale. I blew on the whistle when he did and made a fuss of him. I repeated the process a few times, and in that same few minutes, I had him exhaling quickly looking for my response. The most exciting aspect of that session was looking into Ingelosi's eyes and noting his deliberation. I was watching him thinking, trying to work out the game I had invented. We were talking to each other. We were playing. These are the happiest times of my life. There is nothing greater than the honour of playing with animals. I enjoy the privilege immensely and at the same time I take the honour very seriously.

To be a responsible animal trainer requires that we are clear in ourselves. To be in relationship with anything means commitment to giving our best. Having pets to gain companionship or love will not work. If you do not wish to give companionship and love, you will never be open to receiving it.

Loving them enough to lose them

Working with animals does require a level of commitment that can at times be overwhelming. Particularly if you are working with the principles of relationship. Animals die, and usually they have a shorter life span than we do. The death of an animal that you work closely with is the same as experiencing the death of a family member. At the end of the day in a job where you work with animals you don't just turn off the computer and lock the office door. Your attention needs to be potentially available twenty four hours a day. After working with the dolphins, penguins and seals for more than ten years, I felt the need to move on. I moved with my family from South Africa to London. Saying goodbye was difficult, but also somewhat of a relief. I thought that skipping the country would provide me with the space I needed to start afresh and move in a new direction in my life. I was wrong. In a week I was depressed. I remember one evening, sitting in our new home, which was still sans furnishings. I was in mourning and desperately wanting the attention of an animal. Seconds later I heard a scratching at the door. It was a black cat. I sat on the kitchen stair and cried buckets of tears while I stroked my new friend. He visited a few nights in a row. Eventually I went off and tried to purchase a kitten at the nearest pet store, but found that this was a very expensive undertaking in England. Eventually I adopted an orphan 'second hand' cat from the local animal welfare charity shop. Philli became the newest member of the family, and had the unenviable task of filling an enormous gap in my life.

Purpose

I loved living in England. There was nothing that I did not enjoy about London. However, I felt as though I was not fulfilling my purpose there. At one point an advertisement that had been filmed in the dolphin facility where I had worked, was flighted on UK television. The dolphins that I had become so close to appeared on the screen daily, and every time, left me in a puddle of tears. After six months in London, I made a decision to return. My family would have to be uprooted to return to South Africa once more.

When I am working with animals I feel as though I am where I belong. Since my return, I have broadened my perspective significantly. I have spent more time working with other species of animal, and am excited at the prospect of teaching people about working with animals. It is as though, I am expressing my essence when I am doing this work. It is my choice to do this, and my heart sings when I do it. I strongly believe that this is the mindset that is essential, if you wish to succeed at creating a productive relationship with the animals you train. I have worked with many animal trainers, teaching them from scratch, and what I have come to recognise, is that within a day of getting to know a new trainer, I can tell you whether they will make a career of their new vocation. Many people think that working with animals is a glamorous exciting career. The truth is that the honour of working with animals requires a level of commitment that can only be achieved if you truly love what you do. It is not a hobby.

Trainers that wish to train to show the world what good trainers they are may succeed in the short term, but to keep their heart in the job will be impossible. And they will probably move on. Trainers who wish to train in order to be in relationship with animals will always be fulfilled, because the relationship does not require that the trainer shows off to the world or achieves a remarkable behaviour. It requires only, that the relationship remains one of integrity. That the trainer does what needs doing to maintain the animal's total well-being, and that any

Chapter 4 Why we do What we Do

behaviour modification results are simply bonuses along the way. For a trainer of this description, simply being in the presence of the animal is enough. More than enough. Something for which we can be enormously grateful. It is an honour. When we go into training sessions coming from a place of gratitude and humility, everything will simply fall into place.

There was one trainer who began working with us in the dolphin facility. She was well qualified for the position with a degree in psychology and experience showing dogs. She was a beautiful girl, with long blonde hair, slim and well groomed. She would have looked really good on stage. If she ever made it to stage. In her interview she related how she had watched a dolphin show when she was a little girl and that being employed to do what she had seen was a dream come true for her. As it turned out, that dream was short lived. The trainers on stage are only fulfilling a very small function of the job. New staff always have to learn the support duties before they can progress to having animal contact. This is done so that we can ensure that the animals will be meeting someone who will stick around. Relationship is the most important aspect that we develop between animal and trainer. So our new beauty queen had to start work in the fish kitchen, where we prepare the animals' daily food. I walked into the kitchen where she was being taught by one of the other trainers. There she was, with two knives in her hand. One to hold the 'slimy' fish, and the other to try and slice the spines of the slippery thing. She looked at me with shock and horror on her face. Later that day she came into my office, doctoring a couple of broken nails and admitted that she was not cut out for the job. She has become a Soap opera actress since her one day career with us ended.

I watched another young lady try and muster up courage in the face of some of the largest dolphins in any facility. Gambit weighs in at around five hundred

kilograms. When this young lady walked by his exhibit, he noisily chased after her, tail walking and shouting loudly. This is a game he often plays with us. We will turn and chat back to him and sometimes another game will ensue. He was far too intimidating for our newest employee however. In her interview she recounted how she loved dolphins. How they had been a part of her life since she was little. She had posters and books and DVD's of them. But she had never met one. She had a completely different idea of what they were really about. I guess she expected quiet mystical animals. This young lady became an advertising executive. She specialises in advertisements that use animals.

Chapter 5

The Power of Choice – Free Will

We each have the freedom to choose in each moment. We can choose for ourselves, but not for others. When we allow others to choose for us, we withdraw our trust in ourselves to make choices. Accepting the choice another makes for us is our choice.

When we consciously choose to let our Infinite Intelligence guide us, our lives unfold in effortless and magnificent ways. (Arnold Patent, Stephen and Kathleen Norval – Universal Principles)

To maintain ourselves as a positive presence in the lives of those around us, and particularly, the animals we train, we need to ensure that they are with us because they choose to be with us. The choice for the animal must be simple. The thought process would be along the lines of 'I choose to co-operate because the association of the co-operation has a fulfilling connotation.' In other words, if the animal does not choose to co-operate, there can be no aversive repercussion.

We would imagine that this is clear with the technology that we employ. Marine mammal trainers are familiar with the term 'Least Reinforcing Scenario', colloquially referred to as LRS. What this means, is when an animal faults, for example, the standard of a jump behaviour that was requested is not achieved, we do not provide any reaction. Classically, the neutral response is the trainer standing in front of the animal for around three seconds. In human terms, if my son misbehaves, I send him to his room. His room is not sterile and awful. He has a computer which he enjoys. He has toys and books and games in his room. It is not punishment. It is reinforcing for him to be in his room, but it is less reinforcing than having my attention as well. It is his least reinforcing scenario at that point in time.

Its all in your head

It became clear to me, that the LRS concept was not understood by my training team when we were training dolphins to focus, or station calmly in front of us for a period of time. The trainers mentioned that they were confused, as all we were doing was giving the animals an extended LRS. My reply was, 'so what'. But they remained confused. We entered into further discussion on the topic, and I realised that the trainers, used to the ways of the world, where people get fired or disciplined for not complying, were employing the LRS as a form of discipline. In effect, with this attitude, their use of the LRS was becoming discipline, because that was how the trainer was intending the reaction. In fact, when I looked closely at the situation, I realised that I had noted that some of the animals were beginning to look dejected and actually swimming away when they were anticipating an LRS. I had tried to rectify this problem by asking trainers to reward the animals for maintaining focus for the LRS, but this was also not working. I looked at the problem closely and noted that the trainers were using the LRS as a coercive tool.

Chapter 5 The Power of Choice – Free Will

The response was theoretically the same, still the required three second interval, however the underlying feeling was no longer neutral. The intention of the trainer made it different. The trainers believed they were telling the animal they had done something wrong. What they were meant to be saying was that the animal would not be rewarded significantly for their last response. What they were in essence saying was 'You did the wrong thing! The difference was very subtle, but I believe that the animals read it differently. Dolphins, like all animals see things differently than we do, and are extremely sensitive to our body language. They pick up on our attitude. Imagine the attitudes that the variety of trainers could potentially be offering when employing the LRS. I spent time watching the trainers while they were offering the response, and noted very small nuances in the trainer that showed they were not being neutral at all. Slightly clenched jaws, or tightly closed fists. The dolphins were using their survival instincts to read far more into the situation than was meant to be communicated.

When I discussed the application of the LRS with some of the trainers a couple admitted that they felt guilty about the fact that they had to 'discipline' the animal and tell it that it had failed. Others were frustrated with the animal for not complying. A trainer even admitted that they felt like the animals appeared to prefer other trainers to them, and when they had to LRS the animals, they felt that they were making that situation worse. These excuses sound silly, but they are the result of little judgements that we all have in our heads. Sometimes we are even aware of the silliness of how we feel, but remain victim to the lunacy of what the voice is telling us. All these feelings when we are providing the LRS will manifest to some extent in our attitude, posture, body language and communication with the animal. It may be a simple clenched jaw, or sloping of the shoulders, or taut torso. Either way, the communication is no longer an LRS or least reinforcing scenario, which as the statement implies should still be reinforcing, just not as reinforcing as it would be if we were animated and rewarding with primary or secondary reinforcers. We have attached emotions to the scenario, and in effect, it is no longer a neutral response, but holds in it potentially confusing messages. We have made the end product of the required behaviour more important that the relationship between ourselves and the animal; a relationship that should be based on free will.

If the animal chooses not to comply, we cannot take it personally. It is simply the animal's choice. To take it personally, and imagine that it has something to do with us, is pretty egotistical. The ego is a part of all of our characters, and there will probably always be a time when we do take it personally. To be honest with those around us about how we are feeling also allows us the space to recognise these concerns, and so work to overcome the limits they put on our possibilities. My position of seniority with the team of trainers I work with often results in my feelings of 'needing to prove my worth'. My trainers watch me, and if an animal fails, will potentially be judging my actions. I need to remind myself when I am working in a difficult situation with an animal that I am not there to prove anything to anyone. I am there to create relationship with the animals. I have found that it works to be humble enough to laugh at myself when I recognise these feelings. While in session with the animals, this immediately releases the anxiety that is generated by the judgements of what should be happening, and the animals just become characters rather than reflections of my incompetence as my ego would have me believe. In the times when I am at the effect of my ego, everything seems to go pear-shaped.

More than this, as mentioned before, when we become victims of our egos, we potentially enter into a power play with

Chapter 5 The Power of Choice – Free Will

the animal. As a parent I am very familiar with this concept. From my youth, I have a memory of my mother shouting at me when I was being objectionable by refusing to obey her instruction. Cheekily I had asked why I should follow the instruction and in a temper she yelled back, 'because I said so'. I have even found myself in a temper with my children, unconsciously discharging the same demand. You see, it is the natural state of the ego, or the little voice in our head – the 'who I think I am' persona, to fight for its very identity. So, when we are in a calm focussed training session with an animal, and the animal is not complying, it takes a conscious effort for us to remain objective. If we begin to make up a story about why the animal is not cooperating, and end up at the mercy of the story, we are not being objective.

It's not personal

I remember watching a trainer trying to teach a male dolphin called Kelpie not to bite a surf board. The board had until a couple of weeks prior to this session been a toy for him. As a result, Kelpie was obviously battling to understand this new concept of keeping his mouth closed while interacting with the board. He kept putting the board in his mouth. In the games he had had with the board, he had learned how to sink it under the water and lodge it under an overhang. Kind of like the kids who have those toys where they have to fit shapes into the right places in a bigger ball. The trainer had made the common mistake of moving too quickly with the training and not communicating effectively as a result. But the trainer was not seeing it that way at all, and had become very annoyed and was no longer calm. As a result, she was not even thinking clearly, and in retrospect, should have left the session and come back later with a different plan in mind.

What made matters worse was that many junior trainers were watching the session, which clearly left the more senior trainer feeling like she had something to prove. The session became an absolute power play between a trainer who wanted the dolphin to do what she wanted the dolphin to do, and a dolphin that was initially confused. Confusion can become frustration, but fortunately in this instance, it was not the case. Kelpie appeared to be having great fun getting the better of the angry trainer who at this point was offering the dolphin a very novel reaction for not doing what was being requested. The session spiralled out of control, and the dolphin eventually refused to give the board back to the trainer at all. If we had videoed that session and played it back to the trainer, we would have been able to show the trainer that the novel reactions she was offering had actually trained the animal to respond in the manner that he had.

Kelpie did eventually learn how to surf on that board, but it took a year to train him. The trainer that took over the training trained the behaviour really really slowly. It had to be done this way because he was so conditioned, at this point, to chew the board. The association with the board was chewing it. Steps to train the behaviour began by only showing him the board and rewarding him for being calm and not opening his mouth when he saw it. Next step, touching his mouth to the board. These two steps took a month on their own. Next step was the trainer holding the board and moving a few steps so that Kelpie targeted to the board with his mouth closed. This led to the trainer walking next to the edge and moving, letting go of the board for a second and then taking hold of it again. In that entire year that it took to train that behaviour, Kelpie was never given the opportunity to chew the board. Eventually it was so entrenched in him not to chew it, that it was like learning a whole new behaviour. Chances are it could have been trained faster, but the trainer was not prepared to take that chance. And he also learned a fair amount of other behaviours in the same year, so it was not as though this was all he ever got to do.

Chapter 5 The Power of Choice – Free Will

Emotional Consciousness

All our fears and concerns, whether real or imagined, will always play out as narrations by the voice in our heads. It is our responsibility to be conscious of where we are emotionally, and to remain calm and focussed, and remember that we can choose to respond to that voice. We are listening, and we are not that voice. We don't have to listen to that voice as though it is commanding us, because when we do, we become reactive, and our body language and decisions will always tell the animal the real story. They will simply mirror our state of being. If we enter into a power play we are no longer offering the animal the choice to cooperate. The more we pull in one direction, the greater the chance of the animal pulling in the opposing direction. This is because there is no choice for the animal. We are expecting the cooperation on our terms. We are forcing our hand. And even if we do succeed under these circumstances, the behaviour that we train as a result will not be as strong as it would be had the animal had the free will to make that choice. This is most easily understood when we look at punishment in its rawest sense.

Choice is power

When trying to put a horse into a horse box, the grooms and trainers coerce it into the confines of the box by pushing it into the enclosure, taking away its choice by promoting forward movement with ropes behind it and whipping it with a crop when it tries to get away. The horse is boxed without choosing to co-operate. In fact, the association the horse has with the horse box is probably worse as a result of the fuss and bother. The fuss did achieve the end of boxing the horse, but the horse may not willingly make that choice again unless the same or probably more coercion tactics are used in the future. Horses have been successfully trained to walk into the box, and rewarded for their cooperation using more suitable training technology. A woman from a horse rescue centre called me frantically one Tuesday morning. She had a horse that had been at a competition on the weekend before, and the same horse was due to compete the following weekend. At the previous weekend's competition, the horse had become uncontrollable when they tried to put it in the horse box to bring it home. She described the scenario and was close to tears as she mentioned the crowd of people that gathered to watch, all offering advice and ropes and edging her and her groom on in an effort to get the horse to go into the trailer. With all the pressure, the animal eventually reared so frantically that it fell on its back at one. Eventually a vet was called to give it a tranquilizer so that they could take it on without injuring the animal or themselves. I was concerned about the short notice, as I prefer to be able to teach the animals slowly, however I took a ride out to her farm and worked with the horse. He was really smart, and she was amazed at how easily he learned. Within fifteen minutes he was choosing to walk into the box with me. We did not close the trailer that day, and I calmly backed him out every time he took a step further into the space. I could not return that week, but carefully explained to the lady what to do. She listened and carried on with the communication system I showed her, and had no problems going to her competition the following Saturday.

These horses are more likely to cooperate in the future because they were the operators in the procedure. They made a decision, a clear choice, to cooperate. This concept cannot be difficult for us to understand. As human beings, we are more confident and cooperative if we are able to choose to cooperate. When we are forced against our will to cooperate, we usually rebel, or are very miserable with the concept of the coercion. If we are given the option at a future date, not to co-operate, we will choose that option, particularly if there is something more motivating to do. My son was told a million times not to play with electricity. He was nine years old

Chapter 5 The Power of Choice – Free Will

and quietly playing in his bedroom. I was in the study on my computer when suddenly I heard a yelp, and the electricity tripped. My frustration at the lost work I had done quickly turned to concern. I rushed to his bedroom, and he was standing in the corner near the plug point, with his hands behind his back and looking very pale. I asked him what he had in his hands. Sheepishly he put them out for me to see. They were black from the shock he had just given himself. He has not played with electricity again.

Seals that we work with are all required to be crate trained. If we need to move them to another area for any reason, say a fire or bomb scare or flood, we can do so in the crates. The seals are also trained to kennel into their own one metre square kennel into which they are required to enter at the start of each session. There are at least four sessions a day, and seven seals. If we had to coerce them into the kennels or crates, it would be most unproductive and stressful for trainers and animals. To train a seal to kennel or crate, you never shut the door behind it if it is trying to leave the space. If it makes a move to exit, it is telling you that it wishes to leave. Their choice. Their wish is our command. So you have to let them go. And if they are calm and allow us to shut the gate, they are given a good reward of fish. Their choice.

Their kennelling and crating behaviour is certainly more consistent this way. Consistency in their training has actually resulted in the kennel spaces being very motivating areas for the animals. When working with newer seals, we are always conscious of where we are in relation to the water. The water is the seal's safety zone. So, if they wish to leave the session, or get a fright, they will head for the water. Hlabathi (Zulu word meaning earth), is a fairly skittish little seal, but her trainer has worked wonders achieving her increased confidence. The trainer was working with Hlabathi, and there was a loud crash from close by when some contractors who were working nearby dropped some heavy equipment. Hlabathi startled and headed straight back to her kennel, a place where she obviously felt comfortable. She had the choice to go to the water, but chose her new safe place instead.

Working with horses in the round pen often yields similar results, and very often confuses trainers who are not expecting the horse's response. A trainer was working with police horses, and decided to desensitise the horse to a cool drink bottle filled with stones. The horse had a saddle on, and the trainer was taking the animal through some ground work at the time. The trainer made the mistake of tying the bottle onto the saddle without introducing the animal to the noise that the bottle made. The trainer then stood in the middle of the round pen and asked the horse to move around. The horse obliged willingly. The horse understood the saddle, but this time round, when he moved he heard the rattle, and became anxious. The horse had a relationship with the trainer, and obviously recognised him as the leader, so without hesitating, trotted towards the trainer directly. The trainer misinterpreted the action of the horse, thinking the horse wanted to attack him. He was wrong. The horse wanted his help. The trainer rushed out of the round pen in fright, and it took the onlookers to explain to him what had actually occurred.

Free will in our relationships with animals has one more important implication. If we are comfortable that there is choice, and confident that we are a motivating force, the comfortable choice that the animal makes will be the natural choice, in alignment with our chosen method of training.

Who are we?

As this book progresses, it will be helpful for you to understand the difference between who we are and who we think we are. To help understand, it may be easier to name these two states of being. Who we think we are can be termed as our mind. Who we really are is our heart. This distinction

Chapter 5 The Power of Choice – Free Will

makes sense because if we wish to connect to who we really are, our heart, we need to access our feelings, free of any judgement about those feelings. Also, very often, when we are confused, it is because we have analysed our situation into oblivion and have lost all perspective of what is really occurring – we are stuck in our head. I have experienced and heard many of my friends relate how while battling with a problem, they found that the solution came to them in a flash of insight and a most unexpected moment. An "aha" moment, when we were sleeping or relaxing. This is because the alpha state was achieved, and the limited choices that the beta state offered were forgotten.

Chapter 6

Responsibility – Cause and Effect

Anything we experience is an effect of something we have chosen. We are always at choice if we note that we are the cause. Because there is always something in the effect that we can choose, even if it is the way we feel about it.

We need to remember is that we have free will. We can choose whether we wish to listen to the mind when that voice in our head is rattling off instructions, judgements and demands. With this insight, we have the opportunity to access a much greater sense of wisdom.

No problems!!

As trainers, all we have at our disposal are the training methods and ourselves. Every situation that is presented to us in our training sessions is simply that. In a visit to Jamaica I really enjoyed the rantings of our tour guide when, just before

Chapter 6 Responsibility – Cause and Effect

she sang Bob Marley's 'Don't Worry, About a Thing', she said that in her country, there were no problems, there were only situations. It is helpful to think of training scenarios in the same way. There are no problems. There are only situations. If we look at the situation without judging it, we will look at it rationally, see what needs doing, and simply choose a course of action. We will also tend to keep the responses we have to any particular situation simple. It is when we are in justification mode that the solutions to our training challenges become complicated and confusing – for us and the animals alike. So, now when faced with training challenges, I have a theme song, courtesy of the king of reggae, that plays in my head.

The seal trainers were frustrated with the fact that during their shows, the audience members would very often stand up and walk into the wings of the auditorium in order to try and get a clear photograph of the seals. This situation occurred about once or twice a week, and the sudden movement in a new area seemed to unnerve the seals. All sorts of measures were taken to try and counteract the problem. Chain fences were put up as deterrents to the public to try and keep them from moving to the problematic area, signage was put up, public were warned to keep in their seats during pre-show announcements. However, the situation remained problematic as there was always a foreigner or a child that would break the rule unintentionally. It seemed that the only solution was to redesign the auditorium to prevent the problem from occurring. Eventually the trainers worked to create a solution that was far more creative. They trained the seals to accept people wandering around in the area. Not only did this make the seals more confident on stage, it also led to more creative sections in the show. In fact, the seals eventually began meeting strangers in the show with no problem, an enormous and exciting development to the show.

Anthropomorphism

This is a classic example of how we make excuses for the animals we train, and limit them with our assumptions. In the example, when we addressed the problem by training the seals to accept the intruders, we ended up with more confident animals who were more comfortable performing in their area. This is much more in the interests of animal well-being than trying to protect the animal from a problem that is likely to occur. There are many examples of this. In the animal training world we refer to these problems as anthropomorphism which is basically us attributing human qualities to animals. In other words, we pass judgements on what is happening related to our own experience. Anthropomorphism can be a valuable tool when trying to discern the cause of a problem, but it can be extremely dangerous if used unwisely. In a group of dolphins one animal had died. The trainers were understandably devastated, and very concerned about the animals that were experiencing this loss. They felt that one particular dolphin was in mourning. As a result, she was given loads of attention, and within a month, she was being treated as though she were swimming in egg shells. If she looked sideways for a moment too long the trainers thought she was ailing, and began to do a song and dance around her. Eventually the dolphin was not participating in training sessions. She seemed to have lost her drive. Out of session she lay quietly in a corner. Her eating was erratic too. Medical intervention was imminent. But a last resort was adopted, and a decision was taken to treat her normally. This solved her problem. All it took to turn the situation around was to work her again within in clear boundaries. Within a week she was back to her normal self, vocalising for the trainer's attention and eating properly. All that was wrong was the trainer's overly protective concern. The trainers believed she was ailing and so created an ailing dolphin.

In our relationships with animals, all we have to work with are the training tools at our disposal. We have our intuition,

Chapter 6 Responsibility – Cause and Effect

but it is vital that we draw a distinction between this gut instinct and anthropomorphism. What we feel and what we feel about the situation are two very different concerns. One is our heart telling us a reality, and the other is our mind telling us a story that is based on our judgements and perceptions and experience. We need to be absolutely clear on the distinction. The difference is that the mind creates anthropomorphism, and it generally creates this story when we are not in a training session with the animal in question. The heart, on the other hand creates empathy, and we feel that while in the session. Usually, if we listen to the language of the story we can hear the distinction. The mind usually says things in the following manner. 'I think…' or 'I believe…'. The heart or our true gut instinct language is 'I feel.' But it is a feeling without a judgement. So, it would not be that she feels sad because she lost a calf. In fact the moment we begin to put it into words, we have moved out of intuition and into the mind.

Power Talk

Taking responsibility for our actions is such an enormous lesson that we can learn from animals and take into our lives. Listening to the way in which we conduct our conversations on a daily basis is an excellent measure of whether we are doing just that. Listen to how you speak and what you speak about in your daily small talk conversations. This has certainly taught me a great deal

about myself. For instance, if I catch myself thinking or saying 'it's not fair', I know that I am on the way to blaming someone else for what is going on in my life, rather than doing what needs doing in the scenario I find myself. This photograph is a friend of mine, Dumile, and is taken on Marion Island where this gentleman spent time assisting with consevation research. Dumile is one of the most ardent conservationists I know. He always knew he would do this work, and was prepared to do it for nothing if that is what it took. I first met him when he was handing in his cv at my place of work, and he was humbly telling the receptionist just that.

Language is powerful, because it is an affirmation of what we believe, and that affirmation will entrench what we believe, and move to create our reality. What one person sees and experiences can be completely different to what another person sees and experiences in exactly the same scenario. And we are always at choice about how we wish to be in relationship with a particular experience. Most of us are so eager to be accepted into social scenarios that very often we make the untrue choice.

Let us imagine that we are in discussion with a friend, and they are relating to us about a bad service experience that they had in a supermarket. Because I wish to align myself with this friend, and even though my experience of the supermarket has always been one of good service, I will agree with my friend that life in the supermarket is hard. From here I place myself in the drama of creating a belief system that shopping in this supermarket is difficult. And chances are I will more than likely manifest this experience next time I visit this shop. More than that, I have not empowered my friend to change her attitude by telling her my truth. I have not been responsible in my actions.

Another interesting lesson in our use of language is how we use it when talking to children, and how our use of it in relationship with children creates belief systems in our kids. I have had to stop myself so often from telling my children 'don't do that or you will hurt yourself'. With that statement I am taking away choice from my child. I am telling them that their actions will hurt them. In order to generate a sense of self responsibility in my child, the statement should read 'I would not do that if I were you because it could hurt you'. This takes the predictor autocrat out of the equation and provides my children with the opportunity to make a rational decision. Further to this, what we say to our children or anyone with whom we are in relationship is usually a reflection of what we were taught as a child. Our insight into our personal belief systems is manifest in how we speak. And this personal belief system will illuminate most problems that we have when training animals. Our belief systems need to be consciously chosen and accepted in order that we take responsibility for choices. If we are unconscious of how we think, and ignore the way we feel we are at the mercy of our belief system. We are not taking responsibility for our actions and the way we think. This statement can create confusion in its application, when we are in 'two minds' about something. It seems that our feelings are in conflict. We are not sure how we feel or what is true. All that is occurring in this instance is that we are being confused by fear. If we look at the situation clearly, the truth is always evident. The feeling is the truth. The thought is the lie. Go with the feeling. Interestingly, the feeling is usually the more uncomfortable option, which is why the thought is resisting it.

Belief systems and excuses

This is one of the most interesting aspects of teaching people to be trainers. Belief systems that people hold will always manifest in their relationships with the animals. These manifestations are usually confirmed when I listen to the way people speak, not only about their relationships with animals, but in all aspects of their language. For example, a measure of how successful a trainer will be is how well they listen to any criticism about their sessions from supervisors. A potential candidate for success is the trainer who is open to feedback; the one who is humble enough to listen without trying to furiously defend their actions. They will also be the ones who will apply the relevant feedback in order to improve. In fact, they will seek out feedback so that they can forge their own improvement. The potentially unsuccessful candidate will blame the animal, their supervisors, the weather, or even themselves, in a very attacking or defensive or martyr type mode. They are not open to change. They are playing the victim to circumstance rather than rising high enough to see what is and doing what needs doing. At Sea World, the trainers record their training sessions in a diary for each animal. Reading these records provides insight into how likely a trainer will succeed with a behavior they are teaching an animal. For example, if the trainer writes – the animal is nervous of new people – that trainer has a preconception in relationship with the animal which will make it very difficult for the trainer to succeed when it comes to teaching that animal that it is okay around strangers. If however, the trainer wrote – the animal appears hesitant around new people – then the trainer is not creating a story about how the animal feels. Their objective evaluation provides them a starting point with regard to what needs to be taught to the animal. How the animal feels is out of our control. How an animal responds is within our sphere of influence.

As trainers, with the tools at our disposal, we can only do one thing at a time. When we look at a situation and recognise what is not working, we need to breathe, feel into what needs doing, and then act. Action is all that is required to change a situation. One step at a time. It is the human condition to try and work at speed and fix everything directly. On the opposite side of the spectrum is the common situation where we are overwhelmed with a problem, and as a result we feel powerless to achieve any action whatsoever, so we retreat into a place of helplessness, and do nothing. In this place, we make excuses, and this I have noted is a primary reason for animal trainers not progressing. The horse that cannot be ridden by men because it appears to not like men. A dolphin that

Chapter 6 Responsibility – Cause and Effect

must not be swum with when the weather is poor because it has bitten trainers in this weather. A dog that must be kept indoors when visitors arrive because it jumps up on them. A seal who must not be touched on the behind because she bites when this happens. A rhino who cannot go into the night quarters because it bashes the door. Of course we must be sensitive to these concerns, but they are simply areas where we need to work to sort out a problem.

If we continue to be a victim to the excuses of why we cannot achieve objectives, we set a precedent, and what seems to occur, is the list of excuses gets longer and longer, then we feel less and less confident, and this will affect the relationship between ourselves and animals. The potential will not be realised. There may be animals that are more suited to one activity than another, but this does not mean that we cannot work positively towards a goal for the animal who is less predisposed towards success in that chosen instant. We just need to look at the situation at hand, look at what needs doing, and act. And vitally, remember that we can only do one thing at a time. In an animal training scenario this is even more important. If we only change one variable at a time, it will be clear to us what works and what does not work. If we change more than one variable, and it works, we will not know what solved the puzzle, or worse still, what caused the problem.

Chapter 7

It's not always as it Seems: Beliefs and Illusions

Our beliefs – both conscious and unconscious – determine our experiences. We can recognise what we believe by noticing the experiences in which we find ourselves.

Our beliefs are illusions; they are not real. By quieting our minds and focussing our awareness on what is real – our Oneness – we move beyond our illusory beliefs and experience the inherent perfection of the Universe. (Arnold Patent, Stephen and Kathleen Norval – Universal Principles)

Voice in your head

We need to discriminate once more between the two states of being within us. The first is the voice in our head. That voice being who we think we are. That voice is prattling away telling us what we believe. These beliefs, as outlined in the previous chapter are largely as a result of our experience. As human beings, when we entered into this world we were whole and complete, and free of the voice inside our head. For many of us our first memory is the birth of our illusion of separation. This is what psychologists refer to as subject object consciousness. When we name and recognise something other than ourselves, we recognise ourselves as separate from and in relationship to the world around us. And this separation becomes our reality. All the religions of the world and modern day quantum physics theories tell us that we are one. One energy, expressing at a variety of frequencies. The illusion of separation is the belief that we are separate from that energy. Yet, it is an important illusion because without it we would simply be, and not experience ourselves. Our minds are what break the stillness of our being. Without them we would just be quiet.

Alpha and Beta mind states

The discrimination between who we are and who we think we are can be further explained if we look at the difference between the alpha and beta mind states. These mind states are accepted in scientific literature and are defined in the dictionary in terms of their physiological characteristics. The alpha state has a distinct mental quality we can learn to cultivate. In the alpha state, sensing is predominant. We are mentally receptive rather than active. When we feel an emotion 'in our gut', rather than thinking about it, alpha is at play. It occurs when we let ourselves feel. Extravagant claims are made for alpha. It is sometimes called a 'super conscious' state … Yet, it is completely ordinary. All of us spend maybe an hour or two in the alpha state every day. It always occurs just before and after sleep. In that moment when you flake out just before sleep after a long day your thoughts drift away. You notice the little aches and pains of the body, the cool sheets. You are just there – in the senses, in the present.

We are likely to be in alpha whenever we lose ourselves in something beautiful or fascinating. In the alpha state, we are not judging or comparing or explaining or trying to understand. We are simply in the moment. Not in the past or in the future. This state is when we are close to experiencing ourselves as who we are. It is the state when the little voice is not defining or directing our experiencing. Time is no longer noticed. I remember painting a T-Shirt for my niece. It required a lot of little details and loads of colour. I was amazed when I checked the clock and noted that six hours had passed. If you had asked me the time I would have imagined that only an hour had gone by. I was in the alpha state. The beta state on the other hand, is when the little voice is active. It is the thinking mind state. It is busy. It is directing our action according to our belief systems which have been cultivated as a result of our experience. So we are in the past or in the future. Very often, we are not even aware of the fact that we have choice about whether we listen to that little voice or not.

We can switch between alpha and beta states very rapidly. It may be useful to think of them being on a sliding scale. At any moment you may be eighty percent in the alpha state and twenty percent in the beta state or vice versa. In the very next moment the ratio could change to seventy percent beta and thirty percent alpha.

To illustrate this, imagine yourself taking a sip of cool water on a very hot day after you have taken some exercise. You are really thirsty and hot. As the cool water hits your lips, you feel the moisture and are totally revived with its intensity. You are enraptured with the sense of gratification, with the coolness in your mouth, the thirst quenching delight and the ease in your dry throat. You are in the alpha state until you switch to a beta state and have a thought or judgement, such as 'Wow this is nice. I was thirsty. I must keep water with me whenever I do exercise'.

Alpha Spiritual State

A personal interpretation of the two states relates to a spiritual awareness. We are given the gift of experiencing our oneness by achieving a physical form. In essence, human beings and animals are spiritual beings in a physical form. In

Chapter 7 It's not always as it Seems: Beliefs and Illusions

a physical form, we are able to experience ourselves because of the relativity of our lives. In essence, if we wish to look at it religiously, when we achieve form, we are gifted with free will. We are at choice. We can be hot or cold, because the polarities exist, and we can choose one or the other. Our soul state is experienced when we are in the alpha state. We are not judging or comparing in this state. We are feeling, and connecting with those soul states around us. The beta state is the description of our experience. It is enabled by our experience of ourselves in the relative physical.

With this description in mind, let us consider a training session where we are in the alpha state. We are relaxed and focussed, easily intuiting what is occurring, but an adjustment is required, because perhaps the animal is losing focus and requiring that we redirect its activity. We switch to the beta state and continue to actively direct the session. In this simple illustration, the beta mind state is simply a function. It is the rational mind helping us direct our energy. It is functioning as a result of us remaining calm and relaxed and focussing sufficiently to access the alpha state. If we are unable to access the alpha state, we are unable to be intuitive. A level of intuition allows us to 'hear' whether the animal is responsive or not. Experienced trainers often refer to it as the gut instinct that some have, and some don't. To allow intuition to occur, it is vital for us to maintain a clear distinction between who we are and who we think we are.

The little voice in your head has the potential to limit your potential. If you are operating out of the voice in your head alone, you are operating out of your beliefs, which are a result of your experience alone. This experience, may for example be telling you that you are not a very good horse trainer. Your experience may have yielded a poor result in the past which has helped to cement this belief into your psyche, and the resulting fear is limiting your potential and ability. If you were not operating out of your 'who you think you are' picture, but simply employing solid training techniques in an appropriate manner, you would effectively be operating out of the alpha mind state, and achieving a successful result that is not limited.

Confidence and the Alpha State

Obviously this is easier said than done. However, it has huge implications when we take the time to apply the principle. I watched a young female trainer work a pair of dolphins. She was simply asking the dolphins to practise some behaviours that they already knew. I noted that the trainer seemed distracted, and the dolphins were not behaving with one hundred percent focus. The trainer was renowned for asking more senior trainers to step in and take over the sessions from her at the drop of a hat. Most of the training team were watching her conduct the session. In the middle of this particular session, two of her peers moved past her around to the other side of the pool.

Completely out of character, and for no apparent reason, the dolphins immediately left her and followed her peers to the far side of the pool. In discussions after the session I asked her how she was feeling during that session. She admitted that she did not feel as competent as the trainers watching her. She also said that she felt distracted, and did not feel comfortable working with the animals with so many people watching her. She felt judged, and 'not good enough'. For me, it was very interesting that the animals had responded exactly as she believed they would. For the trainer, this was a moment in which she saw how powerful her belief system was. She had enough experience at that point in her career to know that her beliefs were not real. She had achieved enough success in other sessions to know, despite her low opinions of herself, that she was competent. Yet, she noted how powerful her beliefs could become if she let them take

Chapter 7 It's not always as it Seems: Beliefs and Illusions

hold of her attitude. For her the lesson was that she needed to address how she brought herself to the session if she wished for success in that session.

Years ago, before we bought him, I was learning to lunge a horse. His name is Mushatu. Mushatu is a thoroughbred, and an ex race horse. At the time of my lunging lesson Mushatu was very comfortable lunging to his left but stiff and uncomfortable going to the right. This was a result of his racing career where all his training was focussed on the direction of the track. I was working with an instructor, and managing well in Mushatu's confident direction. The time came for me to turn the horse and direct him in the opposite direction. I was fully aware of his history, and anxious about how the session would pan out. I could not get him to move forward. Every time I tried to egg him on he simply backed off from me. I became more and more anxious, and Mushatu was getting more and more edgy.

The instructor I was working with took me aside and pointed out that I was making the horse nervous with my hesitant posture. I heard the advice, breathed deeply and entered the lunge ring once more. I stood firm and directed with confidence, and only moments later, Mushatu responded with confidence and co-operated immediately. My inner feelings were the only reason he had not responded before. My belief in myself gave the horse the confidence it required in order to fulfil the objective of the training session. I must point out at this juncture, as an aside, that I no longer believe it is always appropriate to lunge horses. In order to achieve this, we need to put constant pressure on the horse, and to keep the horse moving, we cannot release that pressure. This is not fair on the animal. There are more creative and fulfilling ways to exercise a horse, even if we don't ride it.

A couple of the dolphin trainers reported to me that a dolphin named Khanya was not responding to the spin cue.

Touching Animal Souls

I worked her once or twice and had no problem with her responses. So, I decided to work another dolphin next to her and have one of the troubled trainers work with Khanya next to us so I could observe. I had briefed the trainer to follow my lead. In the middle of the session I asked the trainer to request the spin behaviour. His response was a retorted 'but she will not do it' but he stepped forward nonetheless and requested the behaviour. Khanya did not move. I asked him to station Khanya once more. Without hesitating I stepped forward and asked for the behaviour. She leapt forward and did the behaviour without hesitation. Later that day as the same trainer was taking Khanya out for a session, all I said to the trainer was 'believe'. Khanya did the spin with no hesitation.

I was fairly junior and so was given a seal that did not have much potential to train as a show seal to. His name was Benji. He was the one nobody wished to work with, and the trainers told me he would not walk over the centre stage area and so could not do shows. I asked when last people had tried walking him over the area, and the answer was about a year ago. I spent a fair amount of time getting to know the seal, and noted how enthusiastic he was about learning new behaviours. I taught him to march, lifting one front flipper as I lifted my leg while walking in front of him. He got very excited about this behaviour as well as a follow behind behaviour where he put his snout in the small of my back and pushed me forward. In a couple of weeks he was performing these behaviours over the formidable centre stage area, and he went on to be a legendary show animal. Benji was a gentleman, and it was an honour to get to know him. When I watched him pass on as an elderly seal, it was with enormous gratitude that I remembered all the lessons that he taught me. Beautiful boy.

Chapter 7 It's not always as it Seems: Beliefs and Illusions

Underestimating Animals

We can become superstitious when working with animals. If they are hesitant or fail once, we may be cautious the next time we find ourselves in the same situation. That caution is a result of our belief about what could occur, and it has enormous potential to manifest in what does occur. It leads to us underestimating the animals with which we work. As trainers we are in relationship with the animals to teach them something that we wish them to do. But this does not mean that this is all there is to the relationship. If that was the case, many of us would not be interested in the art. For me, training is a part of the relationship, but very often the most meaningful moments I have with animals are not related to the training at all.

When I am miserable it is as though my dogs and cats know. They are there to offer support in a way that is different from their normal affections. I remember a dog I had as a child. His name was Whiskey. He was a fox terrier. My parents said he was not allowed to sleep on my bed. But he did. He would hop off and hide under the bed if he heard either of them coming into the room, and as soon as they left join me once more on the bed where he would spend the night. Where I went, Whiskey was always close by. This is why I found what happened one day terribly disconcerting.

I must have been about eleven years old at the time. While walking on the road one day, Whiskey literally turned on me, not physically, but he was baring his teeth and barking his head off every time I tried to come close to him. He was standing on the road facing the pavement and I was standing on the pavement, wanting to join him on the road. A car had pulled up alongside us and the person in the driver's seat was also yelling, but I could not hear what they were saying with all the barking. Eventually the driver got out of the car and ran behind me and pulled me away from the pavement edge. With that, my dog stopped barking and came to me. A puff adder was just off the pavement edge. I could not see it, but if I had stepped off the pavement, I would have stood on it. My only thought on the matter is that my dog was trying to keep me away from the snake.

Underestimating the animals is also a great reason why we make less progress than we could at times. We became a home to an adult border collie called Chocolate whose owners could no longer keep him. Chocolate had to fit into a family of four other dogs when he arrived. He was pretty submissive and kept on the outskirts. Stroking him when he first arrived was out of the question. He was very suspicious of us and would not even take treats from my hand. I continued doing the group training sessions with the four other dogs. This took the shape of clicker training. I would try and feed Chocolate treats during these sessions, but it was rare that he would take them. Gradually, however his confidence grew. He was watching closely. One day, despite the fact that this dog had never had a day's training in his life, he decided to take part, and proceeded, without any training to do all the behaviours that the other dogs had taken a good year to master. He had been watching, and did not need any careful instruction. He knew the drill. I kind of felt that he had a good chuckle that day. I felt humbled at his participation.

Chapter 8

Intuition and Feeling

It is all an adventure. Every waking, sleeping, happy miserable moment of it, as long as you experience it rather than distract yourself out of it.

What are the Limits of Animal Communication?

A friend had begged me to attend this course and even paid half of my course fee to make sure I joined the forum. I even felt guilty when she told me that if I worked with animals and there was a way to aid my work that I had not investigated, then I was not being as responsible as I could be. I sat in the room, surrounded by people who 'loved animals'. I was sceptical and uncomfortable. Each person was taking a turn to introduce themselves and describe why they were attending the workshop. The common phrase 'because I love animals' rung in my ears. My reasonableness stood as a mask between me and the group. I had been coached my whole career to look for scientific back up for my opinions and there seemed to be nothing logical about the principles I was about to look at in this room. My turn came and I simply said – My name is Gabrielle, and I am sceptical. Fortunately the instructor had a sense of humour and was confident in her own right. She simply laughed and said, 'but you are here, and that is what matters.'

The course was called 'animal communication', and was an introduction to a telepathic way of talking to animals. During the weekend we were introduced to various animals who we 'talked' to, and then we had the owners verify and often validate our answers. I was quite intrigued at how much I got right, and fascinated at the feelings I had when I listened to my intuition. I immediately began to consider how and why my intuition and the intuition of those around me was working. It was also at about this time that I recognised that my scepticism and hesitance at being at the weekend workshop was because I was afraid to be judged by those who disregarded the concept.

Telepathic sixth sense type of work is ridiculed by many, and me being in the business of behaviour modification with all its theoretical insight and critical instruction, I had been trained to be analytical. What I discovered, was that I practise this type of intuitive thought every day. If I have an animal in front of me, I tune into what is possible. If the animal looks high in energy, I will easily ask it to display high energy behaviours, but probably when I am working optimally, I do not ask it to lie calmly unless I have primed it to do so first. These snap decisions are a part of what we do in our training relationships with animals. The theory refers to priming the animal in these instances as an aspect of 'setting the animal up to succeed'. In order to achieve this, we need to feel into what they are capable of doing, and provide them with

Chapter 8 Intuition and Feeling

the opportunity to achieve this result. Just like humans, if an animal achieves success, it is more likely to continue to cooperate. Its confidence will grow, and its relationship with the trainer will be based on a solid confident potential to enjoy the training session interactions. The very nature of the word interaction implies that the action is a common activity that is shared by those in relationship – with the preface of the word being inter. We need to ensure that both parties are always motivated to take part in the activity in order to make the session an interaction. In order to ensure this, we need to feel into what the animal is capable and motivated to do. In other words, we need to be intuitive. It is where the partnership comes into play. When we are listing to our intuition, we are providing the animal with an opportunity to talk to us.

Intuition comes naturally to most of us in our relationships with people. Most of the time we are not even aware that we are using our intuition. Imagine yourself arriving at work in the morning. You meet and greet the same people, but you respond to them differently sometimes, based on how you are feeling, and how you read them.

In the management world, intuitive knowing is well applied when managers note and use people in areas where they are likely to succeed. When training staff, it is fascinating too. For me it is particularly so because I am involved in training people to interact and work with animals. More often than not their failure to succeed in this discipline is related to their own personal dilemmas. For instance, let's imagine a trainer called Jenny has a trust issue, more than likely associated with a difficult childhood. This trust issue is usually directed at colleagues. A good manager may intuit, however, that the trust is actually related to Jenny's lack of self-confidence; a lack of faith in her own abilities. In order for Jenny to progress, she needs to trust herself. It may not be wise, therefore, to manage her aggressively. The better approach may be to inspire Jenny with self worth. Help her take note of her good points and acknowledge and reward her when she is trusting and open. A good leader needs to motivate followers to join in rather than chastise them unproductively for not being motivated. In the case of Jenny, this would take a fair amount of intuition in order to be successful. It means that the manager would need to be in tune with and sensitive to what Jenny is feeling.

A dolphin was injured by another dolphin. She appeared concussed and was not breathing. The trainers had to man handle her to the side and fortunately at that point she began to breathe. Her association with the trainers, and possible pain as a result of the injury meant that for the next couple of weeks she was not very responsive. When we tried to work with her in the same manner as we did before her injury, she would not respond. We had to go back to basics and work at rewarding any co-operation. Gradually we increased the amount of co-operation we expected. This worked well. We set up conceivably possible criteria for her to achieve, and she was able to recuperate and achieve a relationship built on trust. We had to feel our way through the situation. Feel how much she was willing to respond, and ask her as much as we were sure she would deliver. This could only be achieved through the use of intuition.

Successful entrepreneurs will recount over and over how important it is to follow your gut. There is nothing more powerful than an idea whose time has come, and if we are in the realm of understanding this, we will feel what the right thing is to do. No business degree will tell you that it is a good idea to develop mobile phone technology. Somebody had the idea, did what needed doing based on their insight, and success of the venture was the result. It never fails to surprise me how successful this venture is. I am a member of the committee for the Animal Keepers Association of Africa. In the remotest parts of Africa, it has been our experience that not many people have access to computers and thus email through this forum is not always possible. However, everyone has a cell phone.

Chapter 8 Intuition and Feeling

Intuition and Feeling

Intuition and animal care

When working with animals, that gut instinct is vital. During training sessions, if we are successful, we are probably using our intuition anyway, but there are also implications when using it for animal care purposes. People will often account stories of how important their gut feel was when treating an ailing animal. Mel who works with Vervet monkeys in Durban, recounts how when they rescued a monkey she felt sure that the monkey was in trouble because of a small wound behind its ear. The vet disagreed and diagnosed the wound as superficial and insubstantial. A day or two later the monkey developed complications as a result of that wound.

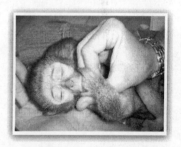

Mel is not a vet, but has a great deal of experience with monkey rehabilitation. She has probably seen more monkeys, and been involved in more treatment and rehabilitation of monkeys than most vets. This experience counts for an enormous amount. She will recognise little things in their body movements and general well-being that nobody else will see. The cumulative amount of experience that people like Mel have will provide them with the ability to arrive at intuitively originated opinions that need to hold an enormous amount of weight in the treatment regime of animals like this monkey. Vets are not in constant contact with the animals the way a handler is. A good vet will listen to the handler. In the zoo keeping industry this scenario is often complicated and can result in a bad ego-based relationship between trainers or keepers and vets. This is because vets have generally been trained to diagnose problems based on the input of the people who attend to those animals. A farmer will call out a vet and give the vet a list of the signs and symptoms they have identified over the recent period. In a zoo type setting, the vet is often employed full time, and expected to have the intuition as well as the medical background.

In my mind, broadly speaking, intuition is the fine tuned ability to read and respond to scenarios. In animal relationships, and particularly in the field of animal training, this implies a greater awareness of the body language of the animals and objective interpretation of our observations based on our experiences and the natural behaviour of the animals. When using intuition in this context in conjunction with consistent behaviour modification techniques, we are guaranteed to improve our success rate.

Without trying to complicate the issue of trainers using their intuition, it is important at this point to note that what also requires investigation in the relationships we have with animals is their perspective of us. It goes without saying that our relationship with the animals is a two way street. We are 'reading' them, but they are also reading us. Their ability to do this is beyond our understanding, and will be further investigated later in this book. But it is vital, to remember that we need to be as conscious and aware as possible of how we are presenting ourselves to animals in order to maintain clear communication. Furthermore, we elaborated in the previous chapter on the subject of alpha and beta mind states. To maximise our ability to be intuitive, our awareness of our state of mind is also tantamount to our success.

So do You have to be a Witch or a Wizard to do this?

My response to the request to attend the 'telepathy' course is testament to my original assumption about intuition and the like. When people talked about quantum physics I could literally hear the fairies playing flute music in the background and the grungy laughter of witches dancing in the moonlight. It all sounded weird and way out there. But

Chapter 8 Intuition and Feeling

I need to recognise that my feelings of scepticism are born out of a life time of training that ensured a frame of mind that is described as logical. So, careful analysis and probing into the concepts are in order. I know I am not alone when it comes to my apparent scepticism. It is even possible that this cynicism has been passed onto me through my education. I remember many arguments that I have had in my life time to attest to this. In my final year of studying philosophy, we had to write a special essay. I wished to write about the potential of the human spirit. My philosophy professors objected and called the subject matter too abstract. I remember feeling very despondent. I studied philosophy so that I could entertain my curiosity of this type of subject matter, and here I was in an environment that I had expected to be free and open, faced with a conservative outlook on my apparent waywardness. That was one of many examples in my life where doors were closed on my intrigue, investigations and questioning. It is the way of the human condition. To teach our societies to conform to the norm, so that we can all get along in the comfort zone. Because the comfort zone is comfortable. It is easy, but sadly, it is often the numb zone, and the zone in which we learn the least about ourselves. And fundamentally, we all strive for the zone of excitement, where we can express every fibre of our being. Potentially we are all at our best, and learn the most about ourselves when we are very uncomfortable. Yet, for safety sake, we strive for comfort. An interesting paradox. But is it only a choice away to keep striving for personal growth.

In terms of acceptable fact, for most of us, if there is not at least a doctoral thesis behind a fact then that fact is not acceptable. Facts, according to the modern world, are not feelings, and can never be determined as a result of our intuition. If we consider veterinary science, for example. I have seen my father when he is consulting with patients. I remember an occasion a lady entered with her miniature Doberman, and when I discussed the case with my father after the woman had left, he told me that he knew what was wrong with the dog the moment pair had entered his consulting room. Yet, he still did all the necessary tests and confirmed his finding before treating the dog. I asked him if this often happened, and he said, yes, nine out of ten cases he was sure, and when he was sure, he was right.

The one of ten times he was not sure, but he admitted that this usually occurred when he was distracted and not functioning in a conducive atmosphere. Although he knew what was wrong with the animals, he still went ahead, as he had been trained by the medical profession, and confirmed the diagnosis with all the necessary tests. Just to stay within the confines of the comfort zone. In a sense, this means that we have been taught to doubt our intuition – as a matter of social conformity. If we lived in America, probably just to ensure he did not get sued.

Yes, even our legal profession insists that we don't base our decisions on our intuition. And frighteningly enough, there are many scenarios where if intuition was not used and medically determined fact was the only case in point, that would have had tragic circumstances. I have two dogs in my family that would have been put down as a result. Tests done on them when they were young puppies told me that one would never walk again and that the other was suffering with an irreversible illness. Both dogs are fine today. The 'cripple' Labrador runs around, not as quickly, but certainly with as much determination as the other dogs. The 'sick dog' is the dominant dog in our family, and the best watch dog in the neighbourhood, and if I take her for a run, she outpaces me and literally and often thankfully, drags me up the hills around our home. I did not agree with the opinion of the vet to have my pets put to sleep. In some place in me I felt that they would make it. And I have enormous gratitude that I went with my gut. The lesson for me is to trust my gut. I don't have anything to lose. As a friend of mine aptly puts it, hearses don't come with trailers.

Chapter 8 Intuition and Feeling

When training animals, even though we may not always admit it, we rely on our gut instincts. This gut instinct is our intuition. Modern day quantum physics theory purports that energetically we are all connected. As an aside, this is not dissimilar to what most world religions refer to with the customary statement 'We are One'. To accept this premise makes intuition and our interrelatedness an acceptable albeit presumably difficult concept to believe. These theories are however researched and theorised by scientists who work in this field. Dr William Braud is a parapsychologist who measured the brain waves of couples of individuals. The pairs being researched were primed to relax together and focus on establishing a deep connection with each other. He found that the brain patterns of the subjects become highly synchronised. Furthermore, the research showed that the most ordered brain pattern prevailed and that a type of 'coherent domain' was established. Energetically, it seems from this type of research, that there is some type of energetic connection between the human subjects that created a scientifically measurable physiological response.

This type of work is echoed by two more experiments that were conducted, and these involved coherence that occurred between animals and people.

Linda Tellington-Jones is the developer of TTouch. This is a method of massaging animals with the objective of helping them to achieve relaxation and healing. Tellington-Jones has patented this method and teaches it to students around the world. Anna Wise, a researcher did some research with Tellington-Jones. Wise simultaneously measured the brainwaves of Tellington-Jones and a horse she was massaging using her technique. Wise used a 'Mind Mirror'. The 'Mind Mirror', is unlike the normal EEG in that it has the unique ability to measure beta, alpha, theta and delta brain waves in both hemispheres of the subject's brain. The 'Mind Mirror' was measuring the brain waves of both the horse and the human in this research. Wise found that as the interaction between human and horse developed, so too did a high level of entrainment begin to occur in the brain patterns of both the horse and trainer. Once again, it seems that the energetic connection exists and manifests physiologically in the subjects.

The late South African vet, researcher, animal behaviourist and keen promoter of animal assisted therapy, Professor Johannes Odendaal conducted an experiment where he measured various indicators of dogs and dog lovers before, during and after they interacted favourably with each other. Blood pressure and neurochemicals considered to be relevant to the state of mind of subjects, such as dopamine, β-endorphin, phenylacetic acid, oxytocin and prolactin, were measured. Results indicated clear tendencies occurred in both species in the same direction and that the tendencies were statistically identical.

This type of research indicates that an inevitable connection exists between ourselves and the animals with which we work. This connection affects both the trainers and the animals that take part in the interactions. As trainers and people who interact with animals, many of us will attest to the fact that the way we feel is affected by our interactions with animals. It is not rare for me to feel lighter and energised after an interaction with them. As a child I was naturally drawn to our pets at home whenever I felt upset, and their presence never failed to lift my spirits. The research is telling us that the animal's state of being is also affected. Because it is clear that the connection exists, our responsibility while in relationship with the animals is suddenly much greater. We need to remain as conscious as possible of the potential affect we have on the sensitive animals with which we interact. Imagine now the trainers of old who went about their work with whips and hooks. Imagine the potential state of being of the animal they were training. Consider the power play that we can find ourselves entertaining during our interactions with animals. What is happening at a physiological and psychological level in both ourselves and the animals during sessions such as this?

Chapter 8 Intuition and Feeling

While teaching a horse training workshop to the police mounted unit, a policewoman successfully made progress with boxing her horse late one afternoon. The following morning she brought her horse down for the next part of the lesson, and began asking him to walk into the horse-box once more. The horse was hesitant and appeared nervous. I was perplexed. Usually the animals make huge progress after some time to think about their prior successes in whatever they have been taught, and when we leave horses on a positive note in the afternoon, we normally find they confidently move into the same exercise when we begin again the next day. I took my attention off the horse and looked at his trainer, and then noted there was something up. I asked her what was wrong. She said nothing and robotically continued trying to ask her horse to move into the box. I told her to stop the exercise and come and talk to me. I asked her if she had had a rough night the evening before. She burst into tears and recounted that her and her partner had had a huge fight that had left her upset. It all made sense. I told her to leave the session and return when she was present and confident. She did, and when she returned, once again made the progress we expected.

To diminish scepticism about intuition, I found it interesting to read the ideas proposed by Malcolm Gladwell in his book Blink. Mr. Gladwell describes intuition as an adaptive unconscious, and 'a kind of giant computer that quickly processes a lot of the data we need in order to keep functioning as human beings. For me, this makes the concept more conceivable. In my experience, there are some people who are more perceptive and sensitive than others. Those with increased awareness appear to be more intuitive.

At this juncture, however, it is interesting to raise the topic of inattentional blindness. As a wife and mother, this concept is often a point of contention in our household. The concept is when we don't see something because we are not open to seeing it. We don't expect to see it, or we focus our attention away from it, consciously or unconsciously. When my child says he cannot find his sports shoes, and I tell him that the shoes are in his closet. He is standing looking into his closet, insisting they are not. I go to the closet, see him staring at the shoes and not seeing them and insisting all the while and as I pick up the shoes and give them to him, he says 'oh, I did not see them.' And he did not. When we brought horses to our property, we had to make sure that there were no poisonous plants in their grazing areas. I had not ever even seen some of the problematic plants in my life before. Little weeds. Now that they have been pointed out to me, I am super aware of them. They jump out at me from a distance, even when wrapped in a mangle of other bushes.

Apparently children that suffer from attention deficit disorders are not as inattentionally blind as the rest of us. My son was diagnosed with this concern at a young age. Before I understood this concept, I would go crazy trying to get him to finish a task. His teacher told me a delightful story of him running a cross country race when he was just eight years old. Being really fit, he won the race, but at the small wonder of all the spectators, because he stopped to investigate everything on his path. He is easily distracted, because he sees things that I don't. His focus is just different, not lacking. He is actually more aware and perceptive than most of us, and yet, this is seen as a problem in the schooling system and in society. We are driven to focus. And in that drive to achieve, we literally become blind to what needs doing done by the deadline. No wonder they call it a 'dead'line.

The reason that this is interesting when we consider intuition is that we are perceiving our environment on a conscious and an unconscious level all the time. This is why subliminal advertising is so powerful. These messages can program people's unconscious to purchase more washing powder or drink more soda; the tools of propaganda campaigns. Our perceptual skills are all being tallied up in our experience, so, when we are suddenly driven by an intuitive thought, it is

Chapter 8 Intuition and Feeling

conceivable that this is a result of our tallied experiences. They are often unexplained in a rational manner because we are not always totally conscious of all our experience. Much of it is unconscious, but it is still filed away.

Mr. Gladwell illustrates this concept well when he relates a story of a fireman chief who was supervising a team who were called to put out a fire in a home. The situation appeared to be a normal kitchen fire. The chief, during the course of the exercise, suddenly experienced what appeared to be a flash of insight, something he later described as a bad feeling, and impulsively called his team of fire fighters out of the home. Moments later the house collapsed in flames. If the team had been still inside the home, tragic consequences could have ensued. When the fire chief was intensively questioned after the incident, it turned out that his vast experience gave him the knowledge required to note that the fire was not behaving as it should. It felt far hotter than a normal kitchen fire for example, and was not responding to the efforts of the firemen in the way that the chief anticipated. The fire was in fact a basement fire, and not a kitchen fire. In the instant of intuition, the fire chief did not stop and process all his experiential data. This would have taken too long, and been grounds for justification and apparent rational judgement to take hold of his feelings. He responded directly to his intuition; a function of his mind, and intuitively made a life-saving snap decision.

Where are the Edges of Our Experience?

While scuba diving with some coral researchers off South Africa's Kwa-Zulu Natal coastline, I was surprised when our boat skipper, on a total whim told us that we were quitting for the day and going home. We were seven miles out to sea in a rubber dingy at the time. The main researcher argued that he still had work to do, but soon stopped arguing when the skipper insisted that he had a bad feeling about the weather. The scientist knew the skipper well and trusted his instincts. We rushed back to shore, and in good time. Gale force winds suddenly appeared out of nowhere, and if we were underwater when they had taken us by surprise, we would have been in serious danger of being lost. It had taken us ten minutes to get to the dive site in the dingy. It took us an hour and a half to get back to shore. If we had been in the water, we would have been in serious trouble. When questioned, the skipper said he just had a bad feeling. However, his years of working on this particular coastline, his understanding of sea and weather conditions meant that the slightest breeze in his face had implications in his experience catalogue. He made a snap decision that kept us out of trouble that day.

So, before continuing, for the sake of common language, let us accept that intuition is the ability to make split second decisions based on experience-founded knowledge that we have. How we achieve all that information and the extent of our sensorial capabilities remains somewhat of a mystery, but it is not necessary to debate this point in order to accept that some form of fine level interpretation of events is occurring when we experience intuition. It is also reasonable to conclude that a fine level of experiencing and perceiving our environment is very much more than we articulate.

To illustrate this with regard to an animal training scenario, let's imagine the following. After working with a dolphin called Jula, in an interaction session, I recount the following. In the session, I was standing next to Jula in shallow water, and was about to ask a female member of the public to enter into the knee deep area next to us. I asked Jula to stay and asked the lady to enter the water. Immediately I felt that Jula became tense. I looked up at the other trainers in the area, but they did not appear worried. There was no apparent or even physical reason for my concern, but I acted on the impulse and asked the lady to sit at the edge for a moment while I assessed the situation. It is important that the dolphin is calm whenever we put people in the water next to them.

Chapter 8 Intuition and Feeling

This is to ensure that the dolphin is enjoying the experience, and that he will associate it with a positive occurrence and be happy to partake in the future. Most important, the dolphin must be calm because there is always the potential for people to get hurt if the dolphin takes fright during the interaction.

Shortly after I asked the lady to wait, Jula backed out of the shallows of his own accord, and swam off. He could not have made it any clearer that he was uncomfortable with the session. Later in the day I discovered that in an area close by, at exactly the time just off shore from the dolphin facility, engineers were busy blasting underwater. There is a chance that explosions were audible as what would have sounded like unusual vibrations to Jula. But the question that I am asked by fellow trainers is how I sensed that Jula was tense, because he did not react in fright, and only backed away from us after I asked the lady to sit down. In retrospect, I have to think hard to answer the question. I imagine that it was just a whim, but then I begin to register the little clues that I may have picked up. At the time of my concern, I was holding Jula's rostrum gently in my hand. He backed away ever so slightly, still maintaining touch, but a little pressure would have been taken off my hand.

My hand in this instance is a target, and if Jula is losing focus, he will break the station of being on target and drift off. This certainly does not happen often, but has occurred once or twice while I have been working with him. This time he did not swim off, which is what would usually occur, but the decrease in pressure was unusual for me, and enough to alert me to the fact that everything was not quite so. Another small cue was the dolphin breathing out slightly harder than he usually did. There was not a marked difference, but once again, enough to alert me to the fact that something was up. The manner in which a dolphin breathes is always indicative of their state of mind. If they are anxious or stressed, their breathing rate will increase significantly. If they are angry or aggressing amongst themselves, they often breathe really hard. The last small clue was not from Jula, but from a flock of local pigeons close by that took off at the same time that Jula breathed slightly harder and did his mini back off.

Pigeons taking flight is something that I have never verbalised as a sign of animals taking fright, but it is an obvious indication that something is happening that affects the flock if they do all take off together. This could have occurred as a result of an under sea blast that occurred offshore – one that I could not hear, but that could have been picked up by the birds and the dolphin. The fact that their taking flight should not influence an opinion that I had while working with a dolphin in a completely different medium would have been arguable if I had stopped and considered my feeling that something was not right in the session. The pigeon action in conjunction with the two other minor variables I have mentioned were enough for my intuition to kick in and make me change my action plan. Once again, however, if I had stopped and thought about the minor variables I would probably, rationally, have overruled them as insignificant. But, I acted on instinct, and as it turned out, my decision was reasonable after all. We don't take in all that we experience on a conscious level, yet, the fact remains that experiences are being tallied up in our minds, and they are playing a part in our intuition. If I had not acted on my instincts, I would have kicked myself and said something to the effect of 'I knew I should have acted'. In light of this if we become more conscious and confident of our intuitive processes, our success rate would increase because we would better set the animal up to succeed.

Some animal trainers rely on their intuition a great deal more than others. Search and recovery dogs for example are exercising their extraordinary sense of smell and hearing in order to accomplish a task. The trainer has to watch the animal and intuit whether that animal is working or not. The trainer cannot tell the dog if it is on the right track or not because the trainer cannot smell what the dog can smell. A friend of mine who researches whales told me about a dog that has

Chapter 8 Intuition and Feeling

been trained to sit on board a whale research vessel and guide scientists to whale faeces. I find it totally amazing that the dog is directing the path of the whole crew of scientists. There is a great deal of money involved in launching this vessel and employing the field staff to carry out this work; but the work is all reliant on the dog directing the events. Relying on the dog to perform the task could result in extra expense if the animal failed. To achieve success, in the process of training this dog, required that the trainer trust the dog to achieve the outcome because the trainer does not have that amazing sense of smell. The trainer had to achieve an understanding of that dog's behaviour so that the trainer could be cued to the possibility that the animal was not attending to the task. This is fine tuned observation. This is intuition in action.

Intuition Improving Communication

When using our intuition effectively our relationship with the animals automatically improves because our communication with them is more effective. That is because effective communication requires that we speak as well as listen. If we listen properly, we will adjust how we speak and thus be more effective in the delivery of what we say. Consider meeting a friend or someone close to you and recognising immediately that the person is out of sorts. You recognise this the instant you see them, and your normal cheery hello is probably going to be delivered in a more empathetic manner as a result of what you feel that person is feeling. This, in part is a result of you reading them intuitively and adjusting your behaviour accordingly. The same type of insight occurs when we begin sessions with the animals. As effective trainers, we will be more successful if we suss the animal out and then pitch the session at the appropriate level depending on what we read into the situation rather than remain conservatively fixated on the task at hand and potentially failing at that task out of our own stubbornness.

Touching Animal Souls

A friend of mine had to do a dog training practical exam. She had just begun work at a training facility where the trainers trained dogs to protect people and their property. Clients would leave their dogs at the facility to be trained in the art of security, and in-house trainers would teach their dogs at the facility over a period of weeks. My friend had to be trained as part of the facility's in house training programme, and had successfully completed the theory and had been mentored in the art of the practical side of things by the owner of the company at the point of her exam. She had been working with one particular dog for two weeks just prior to the exam, and her success with the dog would determine whether she passed the exam or not. Basically she had to do a practical presentation with the dog. As she was about to take the dog, a German Shepherd, out of the kennel and begin the presentation, she noticed that the dog was very excitable.

This was not his normal pre-training session behaviour, and she stepped back and mentioned her concern to the instructor. He told her to stop worrying and to hurry up and begin the session. He flippantly pointed out that it was probably because there was a flock of sheep in the field adjacent to the training ground. My friend was concerned, but trusted the instructor because of his years of experience, and so went about fetching the dog to continue the session. The demonstration would include basic behaviour principles such as heel, sit, stay and recall. It began with her doing lead work with the dog, and at the end of the session she would do the same sequence of behaviours with the dog under her control, but without the lead. During the lead work, the dog participated willingly, although my friend did notice that he was a little distracted, and that he did occasionally look over at the sheep. Just before she took the lead off his collar after having successfully completed the first section of the demonstration, she once again mentioned her concern to the instructor. He dismissed her worries once again and waved his hand. She went back to attending to the dog, made sure she had his

Chapter 8 Intuition and Feeling

attention by asking him to heel and sit once more while on the lead and then went to take the lead off. The dog immediately took off and raced into the field next door to hunt the sheep. Had my friend not been in the exam, she would not have taken him off the lead during that session. The instructor was obviously distracted and not attending to the fine details of the session. He simply wanted the trial of going through a session that he considered to be a task, to be over. This is a classic case of fixating on the end result.

Chapter 9

Developing your Potential to be Intuitive

'Be present as the watcher of your mind – of your thoughts and emotions as well as your reactions in various situations. Be at least as interested in your reactions as in the situation of the person that causes you to react.'
The Power of Now – Eckhart Tolle.

It would be easier for us to predict the behaviour of a human being than an animal because we are more familiar with how human beings think. Or are we? Would anyone predict that a person can become a suicide bomber? Is that conceivable to the majority? Possibly not. However, if we were on close terms with the individuals who took part in these outlandish activities, perhaps we would have an inkling into their potential. It is knowledge borne of the knowing. And the same can be said of our relationships with animals. In my experience, I have seen many trainers become more intuitive as they have spent more time with the animals they train. Intuition is an enormous gift, if we are able to use it. There are methods of developing one's intuitive abilities.

Chapter 9 Developing your Potential to be Intuitive

First Impressions

First impressions do count, but when in a new relationship with a stranger, after we have been introduced, I would potentially be less intuitive about them than if I were in that same relationship with someone more familiar. It is vital to note, at this point, that I am speaking about intuition as it is utilised when we are operating with clarity, out of our alpha mind states. If we are operating with judgements, out of our beta mind state, we will probably have more success being intuitive with strangers. That is because we are still at a point in relationship with that stranger where we have yet to create a whole array of preconceptions and judgements. This array of opinion is what most of us naturally generate when in relationship with people. We categorise those we know according to our experience of them.

But this aside, the familiarity I have with someone will always provide me with insight and information about the individual and so, potentially, this will enhance my assessment of a moment I am sharing with that person. The same can be said of an animal, and more. Relationship generated the correct way yields trust and confidence between animal and trainer. In this well-founded ordered relationship, trainers are more likely to experience intuition. I am much more comfortable with animals I know and I am able to read them more easily than animals I don't know. I have had the pleasure of meeting many animals, and I can honestly say that individually they all have different characters. This is true of all the animals I have met. Seals, dolphins, horses, penguins, ducks, dogs, cats and parrots – all animals, you name it; even turtles and sharks. Trainers of other animals have told me the same stories. I have even watched a couple of fish known as brindle bass being trained, and they each had their own personalities. Their different nuances, body movements, facial expressions, focussing times, etc. all tell their own tale. The more I know of them, through observations and building relationship, the more information I have that will contribute to a sound intuitive assessment.

That said, I will never discount my feelings when I first meet an animal. That first instinct will always be correct. When I think I have been incorrect is when I have taken that first instinct and muddled it up with words and fears. I was asked to work a stallion that belonged to a friend of mine so that I could demonstrate for my friend the way that I choose to work horses. My first instinct when I saw the horse was that he was troubled. I did not let the feeling usurp the training I was about to do, and continued the session. I worked on approaching the stallion on the right hand side, and he had no problem accepting me on this side and letting me stroke him. On the left hand side, however, I experienced a completely different horse. It is not unusual to experience slight differences on either side of the horse, but in this instance I felt that the horse was very perturbed. After I made a little progress near his face on his difficult side, I then left the round pen so that the animal could have some time to reflect on his progress. I went to my friend and asked him if he had any knowledge of why the stallion was sensitive on the right hand side. He recounted how a previous owner had beaten the animal over that side of the face and that the animal had suffered a wound that had taken some time and treatment before it healed. I had recognised that the animal was troubled the moment I met him. His experience with humans had not always been positive. And I needed to build some bridges with him to win back his confidence in human beings. My friend had already established a bond of trust with this animal. But, I was a new stranger, and needed to forge my own foundations with him.

Stuck in the Mind

Having a plan when we enter a session is important, however, getting stuck in the plan amounts to being caught in a power play, which effectively is telling the animal – you will do it my way or else. This is not two-way communication. It is the autocrat demanding an end. If we are open, and flexible in the sessions, we will be in a state of being that allows us to be

Chapter 9 Developing your Potential to be Intuitive

intuitive. It also allows and encourages us to use our intuition to read animals and set them up to succeed. It is all very well going into a session knowing how we wish to train something, but if the plan is not working out how we thought it would, we need to be flexible. When training a dolphin to jump over a water hurdle, the dolphin became fixated on leaving her tail in the water and splashing her body on the spout. This received some reinforcement because as the dolphin, Khanya, did this, she managed to splash me, and not expecting to be wet, I jumped back and yelped. My apparent enthusiasm was reinforcing for her. So, I had to change my plan in mid-session. I taught her to target her tail to a stick with a float on the end. Next I taught her to move in front of me and aim for the target with her tail. I then added in the water spout, and as she jumped, showed her the target for her tail, thus achieving the end of getting her tail out of the water. So, my plan to simply give her the jump cue next to the water spout had to be changed in the middle of the session. Fortunately I had the target on hand, so was able to continue the session, and do something that did not further confuse the issue after my first mistake of creating a fuss about being splashed.

Feeling the Present Moment

I learned a great lesson about myself when I was at a riding lesson. In fact the lesson had been more about ground work with the horses than actual riding. I was leading a thoroughbred through a series of obstacles, such as reversing through some parallel poles, asking him to step on a tarpaulin and to step up on a platform. The horse was being fairly meek with his attempts to cooperate. When the session was over, the instructor asked me how I felt. I launched into a list of excuses on what I thought I had done wrong, and how I could have done it differently if only. My instructor listened to the theoretical debate I was having with myself and smiled. When I was taking a breath and preparing to continue, he simply asked once more, 'How did that feel?' Still I continued my 'coulda, woulda, shoulda, if only speech.' He had to stop me and tell me to listen to the question. In a short retort, I said 'fine', and continued my assessment of the session. He stopped me again, and the words he spoke next always stick in my mind. He slowly elaborated, 'It is never about what could have happened, or what you could have done better. You always need to simply access your feelings. Do you feel that you made progress or not? Do you think the horse understands you or not? Are you confident or fearful? What you are feeling is always the truth. All the other bumph you put on top of it will simply take you away from the truth. That's why, straight after the session, immediately after it and before you begin to judge yourself or the weather or the animal or the tools or whatever other excuse you can begin to develop, ask yourself the all important question. The only important question is *how are you feeling?*'

I knew exactly what my instructor was talking about, but I was so busy trying to be perfect, that I forgot that it is not about me. It is about the relationship that I have with the horse. While I am trying to be perfect, I cannot be perceptive about that relationship, because I am not feeling. I am too busy thinking. I took a deep breath after his words of wisdom, and acknowledged that I felt that the session was better than the previous one we had done. I felt that a little progress had been made, but I was nervous about the platform. My instructor whooped with joy. 'Fantastic,' he said. "Now we have a starting point. Now we know where to work so that we can make more progress next time".

This lesson is powerful. When we begin to become more and more conscious of how we are feeling, we begin to access our intuition. The consciousness of the feeling is the most important aspect of this lesson. And the consciousness must be the truth. The words that we use will tell us if we are considering the truth or not. As long as we can begin the answer with 'I feel', then we are on the right track. On the right track, but not always the track we stay on. In true human

Chapter 9 Developing your Potential to be Intuitive

fashion, we will look for loopholes so that our egos have a chance to express themselves. The rest of the sentence cannot go on to say 'I feel that the animal feels...'. This is dangerous territory. We cannot ever assume what the animal feels. I am not saying that we cannot know what the animal feels, because I am pretty certain that many people are very good at this, but we need to remain in our own feelings in order for our intuition to remain accurate. Our feelings will tell us the truth. Black and white. We enter into grey territory when we begin to base our actions and reactions in relationship to the animals we train based on how we assume they are feeling. This is because when we begin to talk about how the animal is feeling, we are generally looking for an excuse, and in that moment, we are handing the responsibility of the session away. When we remain focussed on how we are feeling, we maintain responsibility for the session. Because, at the end of the day, as effective trainers we cannot change anything else except how we bring ourselves to a session.

The dolphins are taught to lie in a shallow section at the edge of the pool so that the public can have their pictures taken close to the animals. I was about to take a dolphin called Khwezi into a photo session when a trainer who works closely with him told me he did not like doing the photo sessions. As it turned out, the dolphin did not cooperate very well during the session. Every time the people having their picture taken moved around, the dolphin edged out of the shallows. The issue was simple enough. And when I discussed the concern with the trainer after the session, she agreed. Khwezi was fairly new to the sessions, and had not been trained to accept groups of people moving around close to him. This section of his training was stepped up, and in no time he was cooperating perfectly. Who knows how he was feeling. Maybe he simply wanted to back away a little so he could get a better look at this new audience. New people coming close to him might just be 'dolphin television'. For the trainer to simply assume that he does not like it will rob him of stimulation that he may in fact find quite appealing. When we evaluate what is happening, we need to do so with clarity. Because the trainer felt he did not like the sessions, she had also kept him from doing them. So, he was not given the opportunity to become accustomed to his moving television set. Furthermore, the trainer who told me that Khwezi did not like the session was also always tense during these sessions because she had already assumed that the dolphin was uncomfortable. In similar scenarios, with other animals and trainers, I have found that very often, just changing the trainer who is training that particular behaviour can change the animal's attitude towards something. If the trainer is more relaxed and positive, so is the animal – one hundred percent of the time.

Is Your Canvas Coloured or Ready for a New Picture?

Being clear on how we feel is important, not only during and after a session, but probably even more important, in terms of setting ourselves up for success, before the session. If we are low in energy, and not feeling like we usually do, we will go into the session and communicate in a different manner than we normally would. Fear is also a very powerful feeling. It will prevent us from being intuitive at the very least, but also make us present ourselves very differently to the animals. If we are feeling fearful we need to acknowledge that fear, and deal with it before entering a session. If we become anxious during a session, it is better to redirect the session or even end it, so that we can evaluate our situation and then proceed once we have more clarity. This ensures that we are focussed on the task at hand, and allows us to access our intuition and maintain clear communication.

I watched a student on one of the Horse Gentler International clinics try to work with her mare. The mare was an eager student,

Chapter 9 Developing your Potential to be Intuitive

and did everything with a fair amount of energy. I noted that the student seemed a little weary, and in discussions with her later heard the student complain about the horse, calling her difficult and skittish. I had not seen that in the horse, but said nothing. At the end of the day, the student had to lead the horse down a steep road to the horse box. The poor woman was at the end of her tether. She was oozing unconfidence around the horse, and the horse was basically leading her down the hill. The woman was not terribly agile, and kept losing her footing. She was yelling at the horse and pulling on the lead rein, but not communicating anything except chaos to the animal that was becoming quite upset with the whole scenario. The horse was starting to prance around her, trying to avoid the pulling on its head. We stepped in calmly to prevent the situation from escalating out of control.

A fellow trainer calmly asked the lady what she was afraid of. The woman complained that she was going to fall and the horse was going to stand on her. He said he would walk with them and tell her what to do. He first spoke to her a little about something completely unrelated, and eventually had her giggling about an incident that had happened earlier that day. At that point he handed the lead rein to the woman, and then continued the arbitrary conversation as they all walked down the hill together. They were at the horse box without a moment's problem in a matter of minutes. The horse was calm and so was the woman. He then told the woman that she had just successfully led her horse down the hill. He asked her what she did differently. She looked at him, perplexed and told him that she did not know. He pointed out that she had been calm. She smiled, and hopefully took the lesson home with her.

Acknowledge and Exercise

Most of us have had the experience of knowing who is on the telephone as it rings, before we pick up or read the call waiting message; or thinking of someone we have not heard from in a while and then having that person contact us a short time later. These uncanny situations happen again and again, and we always talk about them with awe. There are many different ancient and modern spiritual, philosophical and even scientific explanations why these instances occur. We can theorise about these explanations all we like. The fact is, they do occur, so we might as well play with them. If something happens, such as the phone call turning out to be the person you assumed it would be, simply acknowledge the instance, and once again, most importantly, remember how it felt to be right. The more we access our feelings about this, the more we will begin to recognise what it feels like to be intuitive and then we can begin to listen to our intuition even more. A mentor once told me that I should try this exercise for a day. From the time I wake up, he said I should just listen to my feelings. If they tell me to go a different direction than the one I normally take on my way to work or to drop the children off, I should take it. If they tell me to ignore the phone ringing I should listen. If they tell me to check the dog's foot, I should do so. I must admit, I have yet to do this for a whole day, but there have been occasions when I have listened to that bizarre request that my feelings have inspired me to do, and had the most remarkable things occur.

Recently, on the way home from work, I felt a strong urge to visit a monastery shop on my way home. I responded somewhat hesitantly, and ambled into the shop. I wandered around the shop, literally wondering what I was doing there and as I was about to leave, I heard someone call my name. It was an old friend whom I had not seen or heard from in ages. He was standing next to a pile of books, one of which I recognised. I said hello, and immediately picked up the book and asked him if he had read it. He immediately launched into a description of the awful state of being he was in at that particular moment. He had all sorts of reasons for his concerns, and voiced these too. At the end of the chance meeting, I had not said much at all. What I had said was inconsequential, I thought. But as we parted ways, he said thank you for being

Chapter 9 Developing your Potential to be Intuitive

a sounding board. He admitted that he needed to have someone impartial to talk to so that he could air his views and put them in perspective. He said he felt lighter and in the same breath, said the oddest thing of our encounter. He said 'I was on my way to a friend, and ended up here. I am not even sure why. My friend lives about four intersections back.'

Another good fun story is when a friend of mine emailed me and said he felt sure I would love the book he had just picked up in an airport. The book is one I have already referred to in this book, authored by Malcolm Gladwell and is called 'Blink'. I was not planning any trips to the bookstore, and my budget was fairly limited at the time, so I had no immediate plans to buy the book, but I felt the need to read it as I trusted my friend's opinion. He was emailing me from America, his home, so was not about to pop in and bring his copy along for me to read. I wrote the name of the book down, and made a mental note to keep it in mind should the time come for me to purchase a book. The following week, another friend, who does not know the first, was visiting from London. He came round to my house to have dinner with our family, and had brought me a gift. The book, 'Blink'.

Play with this concept. Easy fun exercises, like taking a deck of cards and guessing if the next one you are going to turn up is red or black. If you think of someone, call them. If you imagine a scenario you want to experience, imagine it in full detail. If you see a goal for you and an animal, feel it. It is fun, and you may even have an adventure you never anticipated. If you find your gut instincts were right, then acknowledge yourself. Your confidence in your ability to be intuitive will grow. You will then be more inclined to respond to your intuition in future scenarios. Some uncanny scenarios present themselves from time to time. Don't be shocked or freaked out. Just play.

Like an ability or a muscle, hearing your inner wisdom is strengthened by doing it. (Robbie Gass)

Again – Never Stop Playing

It is important that we never stop having fun. My niece at age seven had a saying that sums it up – 'If you are not having fun in life there is no point.' If we work with intent, knowing what we wish to achieve, but do so lightly, we are much more predisposed to success. We are in relationship with some incredible animals, and if we are able to play, we will have fun, and they will mirror this fun. When the atmosphere is light-hearted, it is more conducive to us being intuitive. If we are heavy and serious, we are in judgement mode, trying to be right or correct or to produce results. In this state we cannot be intuitive.

Some of my favourite training sessions are the ones in which I had fun, and usually these are the sessions where the greatest achievements are realised. One game we play with dolphins is called creative training. In these sessions, we either wait for the dolphins to come up with something novel and tell them good for any novel response, or we provide them a cue they have never seen before and see what they come up with. This is Gambit's favourite game. He is very creative, and constantly amazes me with the inventions he comes up that we manage to put on cue. In other words, he will show me something, I will put a signal to it, and we will have a new behaviour to add to his repertoire. He does a half a somersault as a result of one of these sessions. He also blows kisses, which is very sweet. He does this by gluing his tongue to the roof of his mouth and then pulling it away to make a sucking sound. He also sticks his tongue out, which is an amusing behaviour to watch.

In a session with two dolphins called Kelpie and Freya, I was working with another trainer, and each of us had a dolphin in front of us. The session to that point had been us asking the dolphins to perform a series of high energy behaviours together. The dolphins were responding with

Chapter 9 Developing your Potential to be Intuitive

great enthusiasm and were eager to participate. On a whim the trainer I was working with asked me to think of a novel cue. I took both my hands and rotated them around each other quickly. She copied the signal and both our dolphins shot off together. In perfect synchronicity and without hesitating, they both did a backward flip. We had never seen or trained this behaviour before. They made it up on a whim, and did it together. I was blown away. They communicated their intent to each other and did it. A real fun session I will never ever forget. Not all the fun sessions end with great behaviours, but they always end up with the fulfilling relationship maintained.

Feedback

By my own admission, I am a control freak. I have to consciously hold my tongue when I am watching a situation. There is a game called the trainers game, which is a very illuminating game to play. One person in the game is the trainer, and another is the animal. The game works best when there is a group of people watching the two at work. The game will begin with the animal person leaving the room. The trainer person and the group then decide what the trainer will be teaching the animal person. The trainer usually has a tool such as a whistle or a clicker which they will use to communicate with the animal person. No talking is allowed. Just the click or whistle. The animal person returns to the room, and the trainer person begins to train that person the behaviour. The animal person is required to be a little creative so that the trainer can blow the whistle or click the clicker or say good when the correct actions towards the behaviour that is going to be achieved are being taken.

I had a group of friends around one evening, and we played this game. I obviously have experience with this concept. None of the friends visiting me have trained animals before, so they were new to the game. We had decided that a girlfriend was going to teach one of the chaps to kiss a statue in my lounge. The session was going fairly well, and the girl had our animal chap in the right area of the room very quickly. Things became a little tricky when the finer details of the behaviour were required. I have played this game many times, and know how crucial the timing of the click or whistle is. I had to literally sit on my hands and force myself not to interfere.

She managed to train the behaviour perfectly. Not the way I would have done it, but it was done. I have found that when I interfere, not only in this game, but when watching people train animals, I hinder any progress. My interference usually only serves to break down the confidence of the trainer, or upset them to the point where they become defensive. Neither results in an intuitive trainer. The bottom line is that I need to let the trainers I watch fail or succeed. All communication needs to occur after the session, and not while the trainer is focussing on the relationship they share with the animal. This enables trainers to remain focussed and in the moment and empowers them to operate more assertively, creatively and intuitively.

He Who Hesitates is Lost

One result of my interference will always be a trainer hesitating. And there is no truer statement than the head of this paragraph. When we are in the moment with the animals we train, we are working harmoniously and in tune with the feedback we get from the animal. When we hesitate and begin to think about what we are doing, deliberating and trying to decide what we should do next, we get into our heads, and this is when things begin to go pear-shaped. There is an important rule for trainers – If you are going to mess up, do it with 100% conviction. Always operate decisively. If you are unsure, do something easy that you know the animal will succeed at doing, and if you are still unsure, end the session. Go away and think about it and come back later.

Chapter 9 Developing your Potential to be Intuitive

This is another reason why it is helpful to have behaviours in the repertoire you train your animal that you can fall back on. I was doing a dog training display for a group of people with my dog Sasha. I wanted her to wave her paw at me while she was standing up. She does this wave behaviour perfectly if she is sitting down.

At one point, Sasha became confused and started offering a whole series of behaviours in a very excited fashion. This could have led to her becoming frustrated and losing focus. She knows what stay means, and this behaviour is well rewarded every time she does it for me. I stopped her mad display by calmly asking her to stay. In this moment, we were both able to focus our thoughts, and begin again on a fresh note after she had succeeded at something she knows well. This meant that we maintained a relationship. If I had become frantic about the fact that she needed to do the wave as I wanted to do it, I would have been concentrating on the behaviour and not the quality of her attention. The quality of her attention is what needed work at that moment. That was the primary step required before I could move on to teach her something new. That is an example of a session that I did correctly. I have many where I have failed to read the signs of what needs attention, and they have resulted in aborted sessions and confused animals.

Chapter 10

Comfort and Discomfort

Fear is a self-limiting illusion.

Learned Helplessness

Imagine being that horse that has been restrained to the point where he no longer believes he can get away. Any horse rider will tell you they have encountered horses like this. Their description of the horses could include any number of the following words – sour, stubborn, meek, obstinate, wilful or disagreeable. They are the type of horses who you don't connect with. You feel guilty riding them, even though they go through their paces. They look miserable. These animals have been 'broken'. They have no sense of self left. They have no spirit. They are not thinking about how to succeed. They are thinking about what they can do to avoid discomfort. They have usually been forced into this place of dumb submission because their people are afraid of them. In the jargon of training, they are the victims of a state of 'learned helplessness'.

They are offered no choice, so zone out of the human world. This horse is not a trustworthy horse. His interests are more important to him than the riders. He is in survival mode, not thinking, just doing the minimum required to be free of discomfort. Our fear has resulted in us limiting our relationship with a horse in this scenario. It has also increased the discomfort this animal feels which affects his quality of life.

Fear versus Love

When we feel our feelings, free from judgement, we are in a place where we are accessing our intuition. When we are not in touch with our feelings, we are avoiding feeling them consciously, and this is usually as a result of limiting thoughts, which very broadly speaking can be termed fear. Great philosophers often speak of fear being the opposite of love. This makes perfect sense to me because true love cannot be felt by the ego. It is not conditional. Fear, on the other hand is a conditional feeling. We are afraid because of something. When we are true in our relationships with animals, we behave confidently. This unconditional confidence is primary in our interactions with the animals in order to achieve holistic success. For me, that confidence is a manifestation of love in action.

It could be said, in the field of animal training, the opposite of confidence is fear; fear of failure, fear of the animal or getting hurt, fear of peer ridicule, fear of not being competent enough or brave enough. So many shapes and forms, and they are always ego based. When riding a horse in a show and the horse fails to perform properly, it may mean a disqualification to the rider. The horse could not care less. When training a dolphin in a show and the dolphin begins mucking about, the trainer may be left standing on the stage looking like a poor trainer, with the dolphin nowhere in sight. The dolphin could not care less. Two important repercussions can result from a fearful trainer that will diminish the success of the training.

Touching Animal Souls

First off, clouded judgement and diminished intuition. While feeling fearful, we remain in a beta state, and don't read and respond as well as we could were we more focussed on the task at hand. The bottom line is that it is not possible for us to remain totally present in the situation. We will be thinking of the fear, and have the voice in our head running at full speed with an unreasonable unjustified nervous commentary. Our feelings will be overwhelmed by the feeling of fear, and it is not possible to feel any other feeling concurrently. Intuition is a feeling state. There will be no place for feeling, and thus feeling the sense of intuition, if we are already filled with fear.

We were working together to start Gandalph under saddle. Wayne was guiding me along. Fortunately, because I was excited at the prospect, yet apprehensive at the possibility of being bucked off. The first time I mounted his back was exhilarating to the say the least, however, I would not have been able to be logical for a moment. Wayne was standing on the ground guiding my every move. Gandalph had no halter on, just a saddle. He behaved like a perfect gentlemen. I climbed off the beautiful horse with shaking legs, but was filled with a gratitude that was overwhelming. Hugs for the horse and for Wayne were the order of the day as tears poured down my cheeks.

Secondly, when fearful, we will be sending confusing messages. Our fear will potentially translate into all sorts of manifestations in our body, and as a result, our communication to the animal will change. Never forget that fear is a physical feeling. Whenever we feel something, there is a chemical reaction in our body. Most of the animals with which we are dealing are well tuned in to body language. Many have an excellent sense of smell, and most also have extremely good hearing. Human beings are pretty expressive with their body language. Even if we try, tell tale signs are visible that a trained human eye can interpret. Guaranteed, an animal has a trained eye. Not only because this is a natural survival prerequisite, but also because in their relationship with us, they have

Chapter 10 Comfort and Discomfort

been trained to respond to our body cues. We think they are only responding to the cue we are training them to respond to, however, very often there is far more at play than we are considering. We had to teach the seals to respond to hand signals as opposed to vocal cues when we began talking on the microphone at the same time as presenting them in shows. As trainers we felt we were taking on a mammoth task. That was not the case at all. What we recognised is that we had small hand cues and body language cues already, and that the seals had been watching these all along. It took a week to teach them to take the signals without hearing the vocal cue.

As an illustration of how we often misinterpret what is occurring, I remember a trainer was trying to teach me how to get her dog to do a section of an agility course. She was pregnant at the time, and battling to keep the energy going during the course. She had taught Beaver to do the entire course, and simply wanted me to help her speed up one section in the course. The section included a tunnel and three jumps after which Beaver would double back for the final section that ended up next to her. She said I just had to run with the dog and he would go through his paces. She was battling to show me the section because of her cumbersome size, and the dog was not responding to my cues. He would miss the tunnel out and then do the last of the three jumps and then head towards her. In a break, so that she could catch her breath, the three of us sat on the grass and discussed the task. She was frustrated, and I could tell that Beaver was getting a little disheartened too. While discussing the subject, we considered what was going wrong. Beaver usually loved the section I was working on. We had already considered the fact that he may only do it as part of a routine, but even when we tried the entire section, he failed to enter the tunnel. Finally, in a flurry she decided to do her best to get him to do the entire course so I could watch her and see what may be missing in my cueing. I did, and had to laugh at what I saw. What was missing was her voice of encouragement. She was extremely vocal just before the tunnel. I took Beaver through the tunnel and over jumps at my next trial. No problem. He was not being cued by her body language at all. He was being cued by her voice and words of encouragement.

Our cueing can be much more subtle than this. If we feel something, at some level, we are showing our feelings. This consideration is born out in an experiment that Mr. Gladwell describes in his book 'Blink'. The research project involved a simple gambling game where the gambler has to choose cards from decks, but does not know at the start of the game that the red deck of cards is the deck where you win and lose a great deal, and the blue deck are the cards that offer a small steady payout. During this game, after about fifty cards, the average person articulates a hunch that they prefer the blue cards to the red cards. After 80 cards most people have figured out the game. Scientists hooked each gambler up to a machine that measured the activity of the sweat glands below the skin in the palms of their hands. Like most of our sweat glands, those in our palms respond to stress as well as temperature – which is why we get clammy hands when we are nervous. What the scientist found is that gamblers started generating stress responses to the red decks by the tenth card, forty cards before they had a hunch about what was wrong with those two decks. More importantly, right around the time their palms started sweating, their behaviour began to change as well. They started favouring the blue cards and taking fewer and fewer cards from the red decks. In other words, the gamblers figured the game out before they realised they had figured the game out. This example is a great one to describe intuition at play, but is added here to note that our feelings are manifest in the physical way before we vocalise them.

On the subject of fear, it is important to realise that it is natural to feel fear. Sometimes we may even be in an intuitive state when a feeling of fear overtakes us and makes us feel uncomfortable in the scenario. It does not matter if we

Chapter 10 Comfort and Discomfort

are right or wrong. Our feeling of discomfort may be one that is telling us we are in trouble and about to be attacked by a dolphin, thrown by a horse or bitten by a parrot. What is important in every scenario where we are feeling fearful is that we do respond because we are conscious of how we are feeling. Because whether we like it or not, our fears are being communicated, and even if that animal was not about to aggress towards us, it may very well take the opportunity to aggress now that we are communicating that we are fearful. Something that they may be reading as a threat, for argument's sake.

Discomfort is a Message

A feeling of discomfort is something for which we need to be grateful, because if we can feel uncomfortable, we know that our feeling of comfort is not a numb feeling. Being out of our comfort zone is a reminder that when we are feeling comfortable, we are not stuck in an unfeeling state. So, gratitude for that feeling of comfort is also a wise feeling to illicit in ourselves. Gratitude is easy when we work with animals, if we don't get into a space where we begin to take them for granted. Taking stock of what we are grateful for is a worthwhile exercise. I keep a gratitude book. It is a list of all that I am grateful for, and I add new items to the list at least once a week. When I am in a poor state, I will read through the list. The list not only includes material objects, but also situations in which I have found myself and experiences I have had. It assists to create a habit out of gratitude.

With regard to fear in the rational state, it is the ultimate in beta mind state. When we are in a fearful state, our fear is based on our experiences and our past. It is a state induced by our history. It has nothing to do with what we are experiencing at the present moment. One may argue that fear is healthy, but in this rational state, it is not healthy. We don't need fear from placing our hands on a burning hot plate. We have knowledge that will provide us with the insight not to engage in this activity. Fear is an anxiety, and has nothing to do with how we use the knowledge of the hot plate having a burning effect. If however we are suddenly overcome by a feeling of discomfort that we may call fear, and this is not a rational consequence, then perhaps we do need to pay attention to the feeling. It may be providing us with intuitive information about the moment. But to simply say, no, I am scared to do that is a historically based beta state fear.

It is not possible to be in control of our feelings. However, we need to be aware of them so that we can understand what we are communicating and how we are being received. Here are some helpful ways to manage fear.

Fear is a messenger. If we let it message us rather than control us, it is our friend.

Be Open

While in relationship with animals and people, we need to communicate openly. It is my experience working with a group of trainers, that if I am in a closed off mood, and not communicative, our training on that day will not be as productive as on days when I am more open. If I am not open, I am not communicating, and I am not providing people around me with the space required for them to communicate with me in an open manner. When I communicate openly, those around me mirror my efforts. I was put in a position of responsibility when I was fairly young and inexperienced. As a result, sometimes in total fear, I worked with an iron fist to ensure that the animals in my care were looked after. This iron fist was part of a suit of armour. I did not show any of my feelings and I shut people out. I can honestly say, that at this point in my career, I experienced the least amount of progress as an animal trainer. It was only when I began to let down

Chapter 10 Comfort and Discomfort

my guard that I began to see that what was required to make progress was a measure of transparency. I needed to be real so that those around me had the space to be real. This is also a much more pleasant environment in which to work and a much easier way to carry myself. It is not only necessary to be open when working with a team of trainers. If we are able to be open in our lives, open to feedback, communicative, and clear on where we are in ourselves, we will take less baggage into training sessions and this will mean we are more inclined to be transparent and clear when working with animals.

I had a trainer in my team who was an ambitious person in her personal life. She had excelled in her sporting career and was well on the way to excelling academically in her private time when she was a part of the team. Whenever an animal did not succeed in sessions, or if it behaved contrary to the training plan this trainer had mapped out, she would take the situation very personally, and I would need to spend time with her explaining away her anxiety and helping her realise that setbacks are a natural part of learning. This young lady worked hard and made excellent progress in her training, but it was clear at times that pressure to perform was overwhelming the situation. The motivation to be in relationship with the animal was lacking. She was too hard on herself to enjoy the journey to the destination. This caused anxiety in her, and at times this anxiety transferred to the animals.

One of the dolphins she was working with at the time had a history of aggression. The dolphin would launch up at trainers without much warning, and bite or knock them. This aggression was potentially very dangerous, and when we tried to analyse what was causing these erratic outbursts, we were not clear on the reason. A fair amount of blame was put on the animal's past, but the fact remained that the incidents did not seem to be getting any fewer. Despite the incidents, there was great progress in the rest of the training. The young lady eventually left to further her academic career, and at the time of her leaving, probably due to a fairly close bond that she and the dolphin shared, we found an increase in the amount of incidents. At this point, a new very calm trainer took over the primary training of this particular animal. What we found after a couple of months was that the amount of outbursts began to fade and the intensity of the attacks also decreased. In my opinion, there is a possibility that the trainer's apparent anxiety-induced need to succeed may have rubbed off on her protégé.

Identify Fears

A senior trainer on my team had years of experience, and was a very creative and successful trainer. Because of her success, she was placed in a senior position in charge of a group of trainers. At this point, progress came to a grinding halt, and personal relationships between the trainers began to break down. Even worse, some aggression was being displayed by the dolphins. When the situation was investigated, it turned out that the trainer was not confident in her role as a supervisor. She felt judged by the juniors in her team. She felt that they were constantly watching her and waiting for her to fail. She became defensive as a supervisor and aggressive as a trainer. Out of humility and love for the animals, she eventually asked for a demotion so that she could begin to progress as a trainer once more. She recognised her fears and dealt with them to ensure her productivity with the animals. A trainer who is fearful will constantly be concerning themselves with that fear or will be trying to succeed for the wrong reasons. They will thus be preoccupied while interacting with animals and not have the gut instinct of a person who is confident of their abilities. Identifying the fears and doing something to rectify them will help the trainer and the animals.

Chapter 10 Comfort and Discomfort

Stay at your Pace

People who are fearful must not be ridiculed. Fears are real to the person experiencing them, and will only retreat from their consciousness when they are acknowledged and accepted. The person then has to make a choice to work through them or simply let them be. I have been put in many situations where I am afraid of an animal when I first begin interacting with it. The only time I succeed, is when I operate at my own pace while creating a relationship with that animal. In some of these cases, it has often become clear that the animal is ready to move onto the next step before I am ready. But I don't go to the next step until I am confident. If I went at the pace of expectation, I would not be doing it with clear confidence that I would succeed. When my relationship – from my perspective – is strong enough, then I take the next step.

Build the foundations first, and those foundations are always <u>my feelings.</u> In another sense, this means that the relationship with the animal needs to be built first. Get to know them before you begin to train them complex behaviours. This is so very true for me. I have never made huge progress with an animal until I have become confident in my own right about my relationship with the animal I am training. I have seen the same with people who I have seen grow relationships with animals, over and over again. It may take a couple of minutes to create a relationship and confidence, but it may also take a couple of months. All depends on the trainer, the animal, and the situation.

Watch Closely

Observing animals in and out of sessions is strongly encouraged. Out of session we will begin to get a sense of how they usually move in different scenarios. When they get a fright they move this way. When they are excited they move that way. They breathe in this manner when they are acting

in that manner. They vocalise in that way when they act in this way. We will also begin to get a sense of what motivates them. Our horses at home have a clear pecking order out of session. When one of us is in the picture, the pecking order becomes even more pronounced. What we see in session with the two of them is something we understand as a result of observing their daily socialising. There was one horse I worked with that used to snort when he became nervous. This was a wonderfully clear indicator to listen out for in the training sessions. Watching others in relationship with the animals that you train will also provide a great deal of insight. I learned a great deal about a horse that I knew by watching the groom interact with that horse. The relationship they had was a longstanding one. I was new to the equation, and I got a fantastic head start by seeing what was possible.

Confidence Breeds Confidence

Throwing people in the deep end works if those people are ready to swim. Not everybody is. Sometimes when they are asked to swim, they are overcome with fear. Others will stay in the moment and make intuitive choices. But there is a risk here, which would result in setbacks occurring because the people will communicate fearfully. In training teams, theoretical training skills should be developed in trainers and we need to let trainers succeed one small step at a time, thus empowering them with confidence. The first behaviour I ever trained an animal to do was an easy one to teach, and I was guided from the side by a more effective senior trainer. I was so excited when I succeeded, and the feeling is one I will never forget, and one I hasten back to in

Chapter 10 Comfort and Discomfort

my memory whenever I become unconfident. If people are keen for other people to develop, then it will be possible to pair less confident trainers with more confident trainers so that a system of mentorship can be practiced. Guidance is a blessing when we are developing our skills.

In the Beginning

Relationships are built over time. At the start, we need to set the animals up to succeed so that the animal is able to gain a positive association with us. It will also breed a successful relationship if we are able to do what the animal already knows, and work through the basics to begin with. For example, a horse trainer I know believes that the first thing a horse should be taught is how to move backwards. He reckons that moving backwards is an extension of the stop. If you can stop a horse, you can control it. With the seals, the new trainers teach them how to go into their kennels before they learn anything else. It is the foundation behaviour. And it is a calm response. With the dolphins, new trainers teach them to station. It is a calm controlled response, and one that is well reinforced. If a dolphin or seal is waiting quietly in front of you, they are focussed and attentive, and ready for instruction. One of the first things most dogs are taught to do, is sit for the same reason. This allows for calm well reinforced cues. When getting to know an animal, we are developing a relationship that needs to be based on positive outcomes. So, we ask the animal to do what is easy for it, and then gradually work towards the more difficult behaviours.

As part of Horse Gentlers we taught the police horse trainers. The first course was conducted on their stud farm where the police have over four hundred horses. When they are old enough, they are brought in from the veldt, trained and then deployed to the various mounted units all over South Africa. That first course was an eye-opener for me. There were twelve trainers learning the Horse Gentlers method of training. They all had different backgrounds in horsemanship. What I noticed on that visit is that the horses in training were all pretty unfriendly. If I walked up to the paddock, they would move away to avoid me. They did this for all people.

We conducted our second course at the facility about eight months after the first. The moment we arrived at the farm I felt a changed atmosphere. The horses in training all seemed calmer. I did not say anything. There were a few trainers from the first course who hung in the background while we taught the new students. One of them, a dear man called Deon invited us to dinner during that week. At his home, he recounted how grateful he was for having learned the horse gentling techniques. He said, that in the past when he used to train the horses it would take him up to three months, and that the end product was not guaranteed.

I asked him how he would go about training the horses. He said they would be herded into the stable where they would stay for up to a week. This was the first step after they came directly out of the veldt where they were used to being in the open. He would go in and feed them and when he could get close enough, he would throw a rope on them and hold them still to get a halter on. Once this piece of equipment was in place, he would take them out to an enclosure where he would hang onto them using the halter and lead rope until they calmed down. He used physical strength and domination to force the horse to submit. He said that this method was not safe and was not always successful. He said that it made him feel uncomfortable imposing himself on the animal. What he loved about our method of training is that it was the first time as a horse trainer where he was getting to fall in love with every horse he trained. His wife giggled in the background in bored amusement.

The relationships that the police are now creating in those formative moments are ones based on respect, choice and empathy. And the results are more agreeable, safer and pleasant horses. Their training is also progressing a lot faster, because the horses are now allowed to think, and choose to cooperate. Injuries suffered by the trainers had also reduced significantly. Deon hit the nail on the head when he said he was no longer working to try and control his fear. He was working effectively, to build a respectful partnership with the horse.

Sing out to be Consistent

A wise sage I know told me that if ever I am depressed, I must go out into the garden and sing my depression to the world. He said I must wail and cry and screech at the top of my voice dictating the extent of my misery. He explained that until I have given sufficient expression to that feeling, I will not be able to let it go. I have tried this a couple of times. Not in my garden, but in the seclusion of my car on the highway. And it works. You cannot stay miserable for long if you are singing about it. When you act something with enough passion, you will become it. So, if we smile and laugh with enough passion, we will become happy. We are chemically wired to respond to the cues that our physical bodies provide us. This is a proven scientific fact. There are even researchers who believe that diseases can be cured using this philosophy. Let us relate this once again to body posture. We need to be conscious of developing consistent body posture when working with animals. We need to go into sessions in the most consistent manner possible, so that we don't communicate confusing messages. And we need to remain aware of the fact that we are in far more control of these postures than we think we are. If we act positively and confidently, we have a greater chance of being positive and confident. Fearful body posture is very easy to note. If we sing that we are afraid, we will be presenting something very different.

Wishy Washy is Diluted Nothingness

To be clear in our communication, we need to know what we are trying to communicate. Have a picture in your mind of the end product and a guideline to follow when you go into a session. This clarity with your intentions will translate in a clear manner of communication. If we remain focussed on an achievable goal, we are less inclined to worry about not achieving a result. So, fear will not result in us becoming hesitant because we don't know what to do next. If we are feeling that fearful hesitance, end it. Then come up with another picture before we continue.

Chapter 11

Listening to the Signs – Mutual Support

Everything affects everything. Watching and listening closely provides us the support we require.

Everything that occurs is an answer to a question we have asked, whether we are asking consciously or not.

I remember when I had my first argument with my son. He was only around three years old. He did not want to put away a toy. He may have had a reason for this, but as soon as he began to object, I flew off into a rage and began shouting to tell him to 'do as he was told'. Other statements that fell out of my mouth without me even having a moment to consider them included 'because I said so', and 'listen to me now or else…'. The argument went nowhere. I went into martyr mode and picked up the toy and stomped off and my confused child ended up in tears seeking consolation from me by following me around. That is when the 'mother-guilt' stepped in and had me reflecting more objectively on what had just happened. I realised that the statements were the stock of comments I had heard adults use as I was growing up. I realised too that I had an unreasonable need to control my son, rather than allow him to express himself.

The most important lesson that day was that there were so many other ways to handle that situation. I imagine that defending our opinions and actions will always be a possibility. I recognise that this same scenario happens to me when training animals. And at times I do really silly things to play out my role as the victimised trainer. But I never achieve success when I am acting out in this manner. And I end up reconsidering the issue, usually in a flash of insight that arrives out of nowhere, most of the time in the middle of the night.

Simple Feedback

As people who are in contact with animals, we have an enormous gift at our disposal. When in a training session or relationship with them, the feedback we receive from the animals – their responses when in session, are all messages to us. Simple messages that we usually complicate with all sorts of nonsense. We spend so much time trying to justify and understand why they are doing what they are doing, that we usually miss the message. It's very simple. If you keep doing what you are doing, you are going to keep getting what you are getting. So, if the animals are not responding in the manner you wish, the only thing that you can change in the relationship, is the way you are acting. You cannot change how they are acting. Sometimes this change requires you to change something in the animal's environment, but more often than not, it is that you need to change something you are doing. All that the animal is doing is providing you with feedback. Their responses are support for your training methods. If you are achieving success, you are doing the appropriate thing. If you

Chapter 11 Listening to the Signs – Mutual Support

are not achieving success, you are receiving support in that you are being communicated to. Something needs changing.

No Wrong or Right, Just 'What Is'

The problem arises when we begin to make excuses for the animal or justify our actions. At this point we are no longer looking at the support. We are denying it and defensively justifying that we are right. The lesson in this is as follows. There is no wrong or right way. When we hear this statement we will let go of our need to be right, and begin to communicate effectively with the animals we train. It is about responsibility. The word responsibility, in its essence is such an important one. It means, literally, the ability to respond. In relationship with animals, if we make excuses about what we are trying to do based on a story we have made up about the animal, we are giving away our ability to respond. The story we are making up is generally an excuse. For example, at a horse training clinic for the police, a trainer recounted a story about a beautiful Friesian called Felix whom the police could no longer use in the opening of parliament because the animal was terrified of flags. My question to the trainer was "What needs doing?" Some of the answers were – find another horse, or tie him to a flag pole till he calms down. These students were new to the training methodology, so I worked with the horse's rider on the 'flag issue'. He first held his horse while I lifted the furled up flag to the animal's face. It was a simple easy movement for the animal to accept. Over the next half hour we worked slowly, lifting the flag up with more and more fabric unfurled, and gradually moving around the animal. We took the flag down every time we lifted it and Felix calmly accepted the object. I left the student with the concept and he took responsibility once again, for what he could do – rather than give the excuse of what he believed the horse was afraid of. This was not really the truth anyway. The truth – as always – was that the horse had not been taught effectively.

I loved watching that flag training progress. In a day the policeman was riding a confident self-assured big black royal-looking Felix bareback and with no bit and proudly carrying an unfurled South African flag. Every time I see success like this I feel so badly for all animals that are forced into situations where they are uncomfortable. When all that is required by us animal trainers is to look at what needs doing, listen to the feedback, and then do it – one confident responsible step at a time.

Tolerating Abuse

Working with animals is potentially dangerous. Some situations are more dangerous than others. For example, there have been many horrific riding accidents, and so, when working with horses, this does remain at the back of our minds, and the inherent resultant fear can result in a break down in communication with the horse and a clouding of our judgement when trying to evaluate concerns we have with that horse. Regal was a cattle horse for many years until his owners were unable to control him, and he was put out to pasture. He remained a good natured animal, and this was impetus for people to get back in the saddle once again. His enthusiastic trainer noted that he was extremely calm until he had a bit in his mouth. When he had a bit, he became uncontrollable.

At this point in this story, it is important to draw on the theory of operant conditioning. Let's reconsider the term counter conditioning. To remind you, it means that we can

Chapter 11 Listening to the Signs – Mutual Support

train an animal to accept a scenario, even if that scenario is aversive, by pairing that scenario with positive reinforcement. An example of this would be rewarding an animal for allowing us to prick it with a needle in order to achieve a blood sample, or teaching a dog that fireworks or thunder are fine. Obviously this sort of training takes time. If we wished to teach a dog that fireworks are not frightening, we would have to desensitise the dog to the noise and smoke one small step at a time. This has been done at a zoo in the states with wolves. It was achieved by using sound effects that were turned up gradually and rewarding the animals for remaining calm and focussed. Over time trainers gradually increased the sound, using a good sound system. When the sound was no longer a problem, they introduced smoke. Then small fire work bangs, and worked up to huge noises, and the final step was a full display. So, the wolves learned to associate the noise and smoke with the positive rewards.

When riding Regal, the bit is an aversive. He was, it appears, in retrospect, running away from the feel of the bit, which had, over time, become acceptable to him through a process of counter conditioning. He knew that if he ran away from the bit, eventually, in an attempt to stop him, the rider would head for home, where he would finally stop, usually in a tight ball of sweat. He had been given the label, because of this, of a horse that was difficult and uncontrollable. But this label did not fit with his character, and it took a perceptive youngster to see this.

His new rider recognised that the bit was the problem. It was clear that he became edgy when this aid was introduced. Desensitising him to the bit did not work. Trying new gentler bits worked to some extent in that he remained calm with the equipment in his mouth, but as soon as his rider applied pressure in an open space, the same flight response ensued. His rider was determined, and eventually took a step that many would not have had the courage to do with such a forward going horse. The steps were cautiously implemented. He began by riding Regal in a lunge ring using only a halter onto which he attached a rein. There was no bit. Regal performed like a superstar. He was attentive and did not try and run off. His rider had ridden him with a bit in the lunge ring before, and found that he was much calmer with just the halter. Before going out into the field, Regal was trained to stop by body pressure, a neck rein and by pulling on his mane and reins. He was wonderfully responsive, and his rider was confident to take the next step. He took him out and the horse performed perfectly, and did not try and run away at all. He even walked the journey back home. Wayne Nicol from HGI did an effective demonstration in the course he offered. He calls up a student and asks the student to step to the right. Wayne stands up close to the student as he makes this request. He asks the student to step forward, then back. Then just as he is about to ask the student to take another step, he pinches the student hard. The student generally does not comply at this point, and sometimes even lashes out at Wayne. Wayne then mentions how difficult it is to think when we are experiencing pain. Very often, the same experience that a horse feels when we are trying to control it with a bit in its mouth or spurs at its side. Horses in these scenarios are working to prevent the pain. They are not thinking.

In the only way he knew how, Regal had tried to communicate that the bit was uncomfortable for him. In fear, the riders labelled him rather than looking at the problem. Regal's response, when carefully analysed, made perfect sense. He was talking. Nobody was listening.

Objective Interface

Communicating with other trainers can be enormously productive. There are trainer forums, organisations and list serves that are available to people all over the world.

Chapter 11 Listening to the Signs – Mutual Support

Sometimes we get so stuck in one way of doing things, or in our own problems, that we cannot see the wood for the trees. It is easy to get bogged down. If we are in relationship with animals for the appropriate reasons, we will never be afraid to ask for another opinion. After all, this would be in the best interests of the animal. And the more opinions we hear the more objective we become. I have many times been watching fairly senior trainers perform training sessions, while standing next to people with minimal experience. The small amount of experience is very often a blessing, because the judgement of these people can be very clear. This has been proven to me over and over. I will ask these youngsters what they see in the session and very often they will see why the senior is having a problem, even though the senior cannot recognise the problem as a result of being too close to the concern.

A session where a female trainer was trying to line a dolphin up parallel to the edge of the pool was not going well. I asked a junior trainer what they saw. The young man replied, "I don't understand what she is trying to do." "Exactly," I retorted, "if you don't understand, you can be sure the dolphin is confused."

It was a hard lesson for me to learn to take feedback, but a liberating lesson when I finally learned it. It is still hard sometimes, but when it is hard, I look at why I am experiencing discomfort, and so the lesson continues. The liberation I experience is the reason I encourage trainers I work with to be open to feedback. This is also a very difficult lesson to teach, and very often I can predict whether a trainer will end up being successful in the profession, simply by gauging how easily they accept or learn to accept criticism. Those who are not open to it usually don't make it. They are the ones who are too caught up in the need to be right. If they take this attitude into a relationship with an animal, they will not succeed. Interestingly, I note that this attitude is usually bred out of a fear of being wrong. This lack of confidence is what is mirrored in the animals, and is not a good foundation on which to build a trusting and cooperative relationship.

SLAM Principle

A mentor of mine, Stephen Norval, taught me an interesting concept which he calls the SLAM principle. It is really simple, but extremely powerful. Basically it stands for "Stop, Look/Listen, Adjust then Move." This is a very powerful technique to use when training animals, and the perfect principle to use if you wish to utilise the mutual support that your relationship with animals offers you. When the training is not proving effective in your desire to achieve an end result, SLAM!!! Stop what you are doing because that is not working. Look at the situation objectively, which may require you listening to more objective opinions. Come to a decision on how to adjust your plan, and then most important, apply the plan – move. If you don't adjust and move, you will be doing what most people in life do – give up. That is not an option. The animals deserve better than that.

The look/listen part of this statement is important. We need to look and listen in terms of taking responsibility. Recall that responsibility is our ability to respond, so we need to assess the situation giving ourselves that ability. As human beings, this is often the hardest thing to do. It is much easier to look for excuses. When working with a seal at the Prague zoo, this was wonderfully illustrated. This work was through an interpreter, so the frustration of the trainers was exacerbated. They could only speak Check, and kept thinking that the interpreter was not explaining what they were saying to me properly. The situation was a seal that at the end of each session would barge out of the gate through which the trainers had to exit. When I asked the trainers what the problem was their retorts were numerous. Here are a few of them.

Chapter 11 Listening to the Signs – Mutual Support

'He is still hungry', 'He is naughty', 'He does not want the session to end', He likes to be with the trainers', 'He knows that the session will carry on if he misbehaves'. After each of the retorts I would ask the interpreter to say to the trainers 'Yes, but what is the problem?' They became very irritated and eventually just went quiet and shrugged their shoulders. I asked the interpreter once again to tell them to tell me what the problem, but asked him to tell them to think about the problem in terms of what they could do about solving it. Very gently one of them answered, 'He will not stay when we open the gate'. 'Yes' I shouted. They were relieved at the simplicity of the answer, but also realised in that moment what needed doing to rectify the concern. The animal needed to be taught how to stay. The rest of the problem solving was easy, once we decided what needed doing.

Chapter 12

They Reflect our Inner State – The Mirror Principle

Everything we see and feel is a reflection of our own state of consciousness. Every person we attract into our lives is showing us some aspect of who we are or who we think we are. Every feeling expressed by another mirrors a feeling deep within us.

This reflection is a gift, for it allows us to be aware of the beliefs we hold. We are then reminded of who we are, and have the choice to reclaim the power we have vested in our beliefs.(Arnold Patent, Stephen and Kathleen Norval – Universal Principles)

A friend of mine Sarah, bought a beautiful four year old palomino. Being the total animal lover she is, she did not meet the horse before the purchase. She went online, found someone selling, bought on hearsay, and paid for the horse to be brought down to our area, from some four hundred

Chapter 12   They Reflect our Inner State – The Mirror Principle

kilometres away. The palomino's name – Dream. Dream is a beautiful horse, but when he arrived, he clearly had his own agenda. He had some obvious problems, and although my friend was told he had been ridden before, it certainly did not look that way. Riding him was out of the question. He was even difficult to groom, and seemed to have a particularly bad relationship with men. Sarah was very concerned. She stabled him with a trainer, and eventually, unable to form a bond with Dream, the trainer admitted defeat. When in session with the trainer, Dream seemed intent on escaping, and became very adept at finding a route of exit. Sarah had made the decision not to use coercive tactics on Dream, and eventually the trainer and her agreed to employ Wayne to advise on the course of action required.

Wayne's first course of action with the four year old was to bring him into a lunge ring out of which Dream could not escape. In the first session they had together, Wayne spent four hours with Dream. Although a very long period, Dream did not seem out of sorts at any point during that session. What occurred in that session was a relationship where it became clear to Dream that there was give and take. When Dream co-operated, some sort of reward was always in order. The reward was in the form of Wayne releasing any pressure. The pressure was not necessarily physical. For example, when trying to approach Dream, if Dream was in flight mode and trying to back away, Wayne would continue to approach. The moment Dream calmed down and faced Wayne, was the moment Wayne would back off. At the end of those four hours Wayne had sat in a saddle on Dream's back. Dream did not break into a sweat during the process, and was kept comfortable with water and hay.

Reflecting our Anxiety

Sarah had to continue the training, which she very aptly did. In a couple of weeks she was backing dream up, walking forward, indirect reining and beginning the lessons of direct reigning. I was watching the session and was proud of the accomplishments my friend had made. Her calm determination had paid off. Then all of a sudden another rider approached Dream a bit quickly, and the horse appeared apprehensive. I watched the next few moments as though they were played out to me in slow motion. Sarah had obviously been caught off guard. She immediately clung to Dream. The horse had been trained to stop with a neck rein and how to disengage his hind quarters, which would have stopped his frightened forward movement, but Sarah did not attempt to use this tool as she was too busy trying to stay on Dream's back. The more she tried to stay on, however, the faster he ran, and eventually he resorted to bucking her off. It seemed clear that in the first few moments of his break from calm, had she responded calmly, and pulled on the neck rein, or moved his hind quarters, that he may very well have been brought into a state of cooperation. But she was too anxious, and he was mirroring this anxiety. When Sarah and I discussed the episode, she agreed. The more anxious she became, the more anxious he became. This is a physical example of the mirror principle, but there are also less physical examples.

I worked with an excellent dolphin trainer fairly early in my career. This woman performed miracles with animals, through consistent application of the theory. She was however, at that point, fairly authoritative, not only with the animals, but also with the people around her. She meant well, and being senior in our team she had an enormous responsibility to ensure the health and well-being of the animals. She had an excellent relationship with most of the dolphins and achieved a measure of success in her training with them that I have yet to see with any other trainer I have worked with. However, Gambit, a large male dolphin in our family did not have a productive relationship with this woman. They clashed often, and their sessions usually ended with one of them swimming or stomping off. I was new to training, and in no position to question her techniques. I did not know why this kept happening then, but I have seen similar instances occur with many people and

Chapter 12 They Reflect our Inner State – The Mirror Principle

animals since then, and it is clear to me now, that what you give is always what you get. The animals in front of us will always manifest the feelings we have. One hundred percent of the time. This is not to say that if we are angry, they will be angry. Or if we are sad, they will be sad. However, they will respond to the feelings that we have, and if we are paying attention, we can use this knowledge to our benefit. If, for example, you are nervous when you are working with a horse, then you need to be extra conscious of not letting this fear control and mess up your training session. Work at your pace to ensure that the exercise is successful.

Children and Confidence

As children, we begin our lives with no preconceptions and concerns. We play in an uninhibited fashion and probably spend the majority of our time in an alpha mind state. When we are sad we cry. And moments later we will be happy and laughing again. I remember noting this while watching my two boys play on the beach when they were still toddlers. They were enjoying the sensations of playing in the sand and at the water edge. If something hurt them they cried. If delighted they giggled, and the changes from one scenario to the next were instantaneous. No judgement, just experience. As adults when we become sad, we look for excuses to be sad. Then we become a victim to that excuse and our sadness. Comments that fly out at the world at moments such as this are, 'you make me so cross', or 'it's not fair.' We blame the world and don't make proper choices because we abdicate responsibility for our lives. We very often don't even cry or allow ourselves to feel that sadness. We become inhibited, and try and fit into a world the way we believe we should fit into this world. There is

no freedom or responsibility here. At this point, we are not choosing to live our lives in this state, but are the victim of how we have been conditioned to live. I remember when I was about eighteen years old I went skiing in Switzerland. I was nervously planking my way down a very gradual slope when a group of toddlers came flying past me, confidently rampaging down the slope at a hectic pace. Their whirlwind energy nearly upended me. They were not afraid, and slow motioned perfectly down the hill, skiing like seasoned professionals. On the other hand, all my well versed and conditioned fears and judgements stood in my way. I took a deep breath when I watched them fly downwards, and threw caution to the wind. I stood like I had been taught earlier that day and sailed down the hill. For a moment or two I had real good fun, till I thought about it again, became frightened, and forgot how to stop. I rolled unceremoniously into a puff ball heap at the bottom of the hill, but took the time to reflect on the fact that it had been fun for a bit and that I had lived to tell the tale.

Our fears and judgements have an enormous impact when we work with animals. The way we hold ourselves in front of animals is reflected in the animals. It is as though without ego, they are simply mirrors of who we are in front of them. My horse riding instructor hit the nail on the head when he observed that confident children naturally manage to use effective body language cues when riding horses. They move without thinking, and this creates all the pressure in the right places. If the kids want to turn right, they look right, swing their right leg back and their left leg forward and against the horse's flank. Their natural weight changes are the perfect pressure cues to tell the horse where to go. My instructor observed that he never has to teach children where to put their legs to cue the horses. I rode as a child and sure enough don't remember being taught these cues. I stopped riding when I was about twelve years old, and resumed again in my adult years. To learn now it is as though the neural pathways have to be burned with a great deal of practice. I spend a great

Chapter 12 They Reflect our Inner State – The Mirror Principle

deal of time in my head, and the cues don't come naturally as a result. However, when I am enjoying the ride, and feeling confident, they do. And my feelings are always mirrored in the horse. When I am having fun, the horse behaves as perfectly as it does when my son is riding it.

Heaviness Rubs Off on Them

There is a great deal of animal therapy occurring in the world. For me, the greatest therapy is to watch the animals I work with, and recognise how I am affecting them. I do this because I care for them, and am willing to be conscious about the results I create in their lives. It is a great responsibility to work with animals, and to do it properly, we need to consider this in detail. The therapeutic gift is when I see how I am affecting the animals around me, it tells me one hundred percent of the time, what is going on inside of me. I watched a lady try and stop her Jack Russell from barking its head off during a dog training class. She smacked it on the butt every time it let off a bark. The dog appeared anxious, watching his owner out of the corner of his eye, and obviously wanting to keep her happy, but also excited about all the activity of other dogs and owners. I quietly approached the lady and asked her how she was feeling. She was distracted in her reply, and as an aside, before she said what she thought I wanted to hear, she bent down and pulled the dog towards her and muttered to him 'Binkly, quiet, you are embarrassing mommy', and then said to me, 'I am looking forward to learning how to train dogs.' Mirror mirror on the wall. They say that owners look like their pets. Maybe it is more apt to say they behave in the same manner.

While on the animal communication course, we had the opportunity of reading a few animals that were brought into the workshop venue. This entailed an owner bringing the animal to the venue, and no person attending the workshop had ever met the animal or owner. One of the pet person couples

I met on the workshop I did was a gentleman and his German Shepherd. The first thing that struck me when the two entered was that it was very obvious that neither of them wished to be there. The dog sat facing the door with his back to our group. The owner may as well have faced in the same direction. His posture was closed off, with his arms folded and his legs crossed in the same direction as the door. During the session we had to intuit answers to various questions that our teacher asked us about the dog, and then the owner would have to verify our answers to those questions. He became more and more relaxed during the session, and his body posture gradually opened up slightly. As he did this, the dog gradually turned around and faced the room and began making eye contact with some of the people in the room.

A seal named Shadow was rescued after he washed up on the beach, victim to a shark attack. His wounds healed quickly, and a decision was taken to keep him in the facility and use him in the educational presentations. His training commenced. The trainer who was given the task of educating him was a high energy woman who worked at a fast pace. Shadow was soon mimicking her movements. He took to the training very well, and it was not long before he had a wonderful repertoire of behaviours under his belt. Unfortunately, he had also developed a fairly aggressive streak, and a point was reached where his trainer actually gave up on him, and suggested he be sent to a zoo because of his apparent aggression. This was clearly a cross road in Shadow's life. A new trainer took over. The second trainer was calm and pragmatic. Within a couple of years the seal had done a turn around, and went on to become a steady, consistent performer.

Animals as Psychics

With experience, and as a result of being a trainer of trainers, I sometimes feel that when you watch the relationships that exist between animals and trainers, you gain an

Chapter 12 They Reflect our Inner State – The Mirror Principle

almost psychic ability to see what is occurring in the trainer's emotional well being. Most of the time, a simple adjustment of that trainer's feelings is all that is required in order for that trainer to achieve success in the training session. Failing to achieve success rarely has anything to do with the animal. Working with animals is a true gift – gratitude is all that is required. When we work from a place of gratitude for this gift, we will adjust when required rather than get stuck in the ineffective power struggle.

Animal therapy is a recognized manner of assisting troubled human beings through designed interactions with animals. This rarely has anything to do with relationships between animals and trainers. I spent time working at a rehabilitation centre for addicts. I taught them about horse trainers. When I arrived there some of them had already been working with the horses for a while, and had made significant progress training them for endurance. What I noted while working with these men was that the art of animal training is true animal therapy. If we care to look at how the animals are responding to us during interactions, we can learn a great deal about ourselves.

It always comes back to the same necessity: go deep enough and there is a bedrock of truth, however hard.

(May Sarton).

Chapter 13

Being Clear – Non-Judgement

We have been taught to evaluate and judge virtually everything that we experience. However, in Universal terms, there is no such thing as good or bad, right or wrong. Everything that occurs is just another event. By judging something, it becomes for us the way we judge it, and stays that way until we release the judgement. Judging anyone or anything as being less than perfect blocks our ability to respond to the essence of the person or thing and creates discomfort within us that can only be relieved by opening our hearts, first to the judgement, then to the person or thing we have judged.

Forgiveness is the willingness to open the energy locked up in our judgements by acknowledging, accepting and appreciating what is, just the way it is. Opening our hearts allows forgiveness to expand into unconditional love. (Arnold Patent, Stephen and Kathleen Norval – Universal Principles)

Fate is the cop-out that we adhere to when we don't wish to take responsibility for our ability to choose.

Chapter 13 Being Clear – Non-Judgement

Doing what needs doing

A rescued South African fur seal called Moya presented us with huge challenges in her rehabilitation process. At first she had to be force fed, and in this process it was not long before she was labelled, understandably, as an aggressive animal. The practice of force-feeding a fur seal is not an easy one. The animal has to be straddled by one person, and this person must ensure that enough pressure is put on her to ensure that on the animal to prevent it from escaping, but she must also not be hurt in the process. This first person must hold the animal's head with both hands to keep it from lunging around and biting people. The neck of these animals is extremely flexible, and they can lunge in any direction to achieve that very nasty bite. The second person will open the mouth of the animal using both hands and at times even bandages are tied to the top and bottom jaw to force the mouth open. The third person puts the food in the seal's mouth. This sounds simple enough, but the animal is wet, and very slippery, so all involved need to be particularly vigilant in order to prevent injury to seal or humans. Moya had bitten four different people to the point where they were no longer interested in the possibility of further injury, and so very hesitant to be involved in the force-feeding procedure. She had learned that people will leave her alone if she bit them. Moya was steadily losing weight. She was receiving enough to sustain her, but the stress was going on for too long, and chances of her suffering with ill health were becoming a possibility. She was also beginning to become fairly reclusive, which is a very bad sign in this process. To ensure Moya's survival, we needed to take urgent alternative action.

We looked at the label of aggression, and all the fear that we had of the process, and had to take careful stock of the situation. In a more objective evaluation of the events, we recognised that our judgement of her was standing in our way of progressing. Three of us were brave enough to make a commitment to the animal and be there every day until she began eating of her own accord. We decided that there was a possibility she was confused, and that we needed to make every effort to be consistent, and to let her know when she was doing the right thing. Until that point, force feeding sessions with her, which were occurring three times a day, were filled with jokes relating who was going to be her next victim and similar. We chose to be completely calm and positive during the sessions. Using this plan, we managed to get her to eat of her accord, in less than a week. We decided that we would communicate with her, telling her she was right by reducing our restraint on her everytime she swallowed or tried to swallow fish. When she began to willingly swallow, we would progress to reducing restraint when she opened her mouth willingly for the food.

In the first couple of sessions, even if it was on the first fish that she swallowed, we let her go immediately. In the next couple of sessions we asked for more, simply reducing our hold on her head if she took the fish in her throat. Because there were only three of us, we could be very clear about making progress, because all three of us knew exactly how well she had done the previous session, and so expected more in the sessions following. We could see her little mind ticking over. The first time she swallowed and we all released her and backed away, she looked at us very superstitiously. We quietly left her quarters and she faced us head on and calmly followed us with her gorgeous brown eyes. The second time, she was a little more animated, and as we left her enclosure, took a step towards us, an excellent sign as till then she had remained backed into the corner. The third time we entered to do a session, she was really easy to catch, possibly anticipating a more positive experience. Every session got better and better. On day four she finally ate. We were all very emotional, and overwhelmed. We had her in mild restraint and she took a fish and swallowed it on her own. I remember wondering if what I had seen had really just happened. My colleague made to stand away, so I realised it must have occurred. The two of us restraining her stood up and backed away from her. She did not notice. The third trainer continued feeding her and she guzzled her fish as though this was something she did every day. She also did not notice that the three of us were all in tears of gratitude and relief. Her well-being was assured.

Our success was as a result of us changing our attitude towards the situation, removing the label from the little seal, and then implementing our plan. Justifications for not succeeding were the reason we were not going to succeed. Determination to succeed, with a clear unclouded plan was required in order to ensure that Moya lived. Judgement limits our ability to act with clarity. Guilt is very often the language of judgement, and simply an excuse not to do what needs doing.

Judge Jury & Executioner – in our Head

Judgement is such a large human condition. It is what we spend our entire lives doing. It's that little or rather, sometimes rather large voice in our head again, that is constantly judging everything around us. That voice becomes more and more eloquent as we learn more and more words, and have more and more experiences. It very often feels as though it is on fast forward. And that voice often prevents us from appreciating the moment for what it is.

That voice is the big old judge, but can be our friend, as long as we remain critical of what it is saying. To be clear on whether the judgment is rational and appropriate requires an open heart. Most of the language of the voice is defensive. 'Do this or else', 'be careful, remember what happened when…'. If we are going to remain defensive, which is usually what the voice determines, we will never get out of our heads, and all the preconceptions that thrive inside us will control our actions and awareness. Preconceptions which are simple judgments based on what we have learned. They are all from the past. They have absolutely nothing to do with the present. One of the biggest contributors to subjective judgement, in my opinion, is language.

Touching Animal Souls

We could use biblical analogy of Eve being tempted by the increase of knowledge and succumbing to that temptation, as the time when language in the form of the spoken word developed. It is the time, in psychological terms when subject object consciousness develops. In quantum physics terms, it is when we decide that we are no longer a part of the whole. It is when the 'illusion of separateness' develops. The word is a remarkable development that aids in defining the subtlety of our experiences. When reading good authors' words, I am often led into a discovery of something I knew about, but given a gift of a more intense appreciation of that knowing. For example, we may know that we love the taste of chocolate, but when someone describes it to us - smooth and rich, melting on your tongue to leave a nutty earthy scent and a pleasurable salivary gland cramp in our mouths, our appreciation of a different aspect of the experience of eating chocolate may develop. Language certainly has it's place, and the power of the void is enormous. To prevent ourselves from falling into the void, we need to use it wisely.

Appreciation of subtlety can be brought into awareness using language, but its development has also given us the ability to fine-tune judgement. Language has made us terribly analytical, and it is a medium that is constantly developing so that we can continually maintain a fine description of what goes on in the world. Uncontrolled judgement will cloud our ability to be intuitive and aware in our relationships with animals. We need to stay in the moment and experience what is occurring for what it is, not what we think it is. Stay in the heart, not in the head.

This probably sounds very impractical. Obviously we need to think, but we need to also trust in our experience and knowledge.

My father is a very experienced vet. He has practised for almost forty years. He has told me tales of how clinically speaking, based on the symptoms two different dogs are

Chapter 13 Being Clear – Non-Judgement

presenting, he would be required to treat the two dogs in the same manner. However, very often he does not. His gut feel, or what he refers to as his gut feel, often tells him to treat the dogs differently, and time after time, he says, his intuition has guided him in the right direction. So, rationally speaking, in this instance, we can note that his experience has furnished him with the know how to make these calls. His practise is simple. In fact, being an old time vet, he  does not have any modern day equipment like ultrasound or even X-ray. If the full clinical was to be done on his patients, more modern vets may reach the same conclusion his gut instinct is telling him. He has relied on his instincts for years, and they are developed. And his experience has developed them. He is not afraid of failing. He is not judging himself in the process. He is simply doing a job he has been trained to do and at which he is very experienced.

If we take this example to the field of animal training, we can see that it can be possible, providing we have some of the right training, sufficient experience, but most of all, confidence in our gut instincts, to make the right feeling choices.

It is important to note that language and our judgemental way of using it cannot replace feeling. No matter how many words we have to detail and describe love, we cannot feel it unless we have had the experience of the feeling. Unfortunately language is often used to replace feeling. We will rather talk about it than feel it. Language is more comfortable than feeling. We forget that being comfortable is the same thing as being numb. And without feeling, we cannot be in a true relationship with anybody, animal or person.

A common question I will get from trainers is 'Was what I did the right thing to do with that animal.' I have illustrated what my mentors have taught me, and my reply is in line with their lessons – I will say, "What do you feel about it? This will often frustrate them enormously, but it is my intention with this question to ensure they take responsibility for their actions rather than explain it all in words, which in my experience they always use at our next debate to justify something they have done which was not appropriate. If they are feeling whether something is working or not, they are learning how best to apply their training knowledge. What I may have told them, will simply become another voice in their heads, which will serve to make them more closed off to the essence of the next session they will experience. By allowing them to feel the answer, they remain open minded, and will be less inclined to get into judgement mode – the language based mode that limits the potential of the next training session they will experience.

I went on an outride with an experienced endurance rider. There were four men in the group all on their own horses. I wished for their feedback on my riding style, and asked for this from them. The endurance rider simply said – "If it feels right it is the right way to ride. The more you ride, the more you will feel what is right. When you ride wrong, its uncomfortable. Get comfortable." In other words, get out of your head and start feeling.

Our obsession with language has also been blamed for our apparent increasing insensitivity to our world and the creatures around us. Nothing can put into words our appreciation of the sunset. When we try to describe the way we feel to determine how we appreciate the beauty we simply step out of the moment of that appreciation, and our rationalisation takes us away from the experience. Without the appreciation of the experience, we lose the experience completely. If we put this concept to work in the relationship we have with animals, it may be easy to describe why we don't communicate with them as well as we could. If we are able to tune into their subtleties without the incessant need to describe the experience, we may find that we are able to read them better. It is imperative that we remain in the moment when we communicate with them, giving them one hundred percent of our conscious attention. This will inspire them with the confidence in our relationship that is required in order for us to succeed in that relationship.

Language Kills the Moment

Language is the judge of our experience. It is what takes us out of the moment of being in relationship with our environment, people and animals, and places us in a frame of mind where we begin to judge, change, plan and define our experience. It is important that we do all these things, but not when we are needing to be in the moment. We can only be effective when we are in the moment. The moment is the only part of our experience that we can affect. It is the only place we have power. We cannot change tomorrow until tomorrow comes. We cannot change yesterday, so we may as well leave it behind.

If, as in the example of the seal Moya, we had continued along a proven track of force feeding in the tried and tested conventional method we had always used in these scenarios, we would probably have lost Moya's life. No two animals are alike. It is required that we remain open to the possibilities and that we don't cloud ourselves with judgement if we are to ensure the well being of the animals in our lives. We need to balance our energy appropriately to the animal before us. This cannot be detailed in a manual. It can only be felt. Always!!

Moya's lesson saves lives

Using what we learned from Moya, a colleague and I were able to make short work of teaching five South American fur seals to eat. The animals were confiscated at the Johannesburg airport after their dealer disappeared. They had not eaten for a while and a sixth animal had already died when the National Zoo stepped in and asked that the animals be brought to the zoo to be treated. They did not know how to eat at this point. Force feeding these animals we used negative reinforcement to teach them to eat. A scientist who works with wild seals and was also a professional at catching the animals was in attendance. We caught them all and held them with a small mattress, and then proceeded to open their mouths and insert food. Every time the seals made an effort to swallow we reduced the pressure on them by letting them go. In no time four of the five animals were eating. On day two of this intervention the one male was still not eating properly, but he was progressing. At one of the feeds on this day he swallowed his last food with people around him, but with no pressure at all on him. I felt confident that he would eat at the next session, so went in confidently. He stood in front of me and looked confused when I offered him a fish. He looked at me and then looked at the mattress that was lying next to us outside the enclosure. We brought the mattress and simply placed it on his back. He ate immediately. We moved the mattress out for his second fish, and he never looked back. There were whoops of delight from all in attendance, and the seal hungrily began satiating his hunger. Five little seals had Moya to thank for their quick eating lesson.

Limiting Potential with Judgement

It is clear that judgement will limit our potential as effective animal trainers. However, there is another equally important part of this equation that needs to be addressed. Being brought up in South Africa in the years of Apartheid presented all sorts of scenarios which I can fortunately appreciate now that the age of tremendous human discrimination is over. I am white, and so was among the 'privileged' minority in this time. I went to a government school, and watched government controlled television, listened to government controlled radio, and lived in a very ordered society where propaganda

Chapter 13 Being Clear – Non-Judgement

was as rife to fear the black man. I was told over and over, from a very young age, and in fact, fabricated scientific evidence made its rounds that black people were simply not as bright as white people. Interestingly, although the apartheid propaganda always left me feeling uncomfortable, there were some elements of the information that I was fed that I have had to consciously work to loose in my demeanour. Primary, was the concept that black people cannot do anything for themselves. This created a sense of pity in me for my darker countrymen and women. Pity, is a judgement. What it does, in effect, is make the person that you pity less than what you are. If you are in a relationship with a person that you pity, you will not trust them to do the job you expect of them. You will be limiting their potential. Fortunately I was able to overcome this silly notion, and today work alongside some black trainers who have extraordinary skills in this field.

Think of this in relation to animal care and training. Very often, because they are so a part of our lives, we anthropomorphically pity the animal for some reason. This pity is a judgement that will serve to confuse the animal and which will limit the potential of that animal. I have often heard the following retort when I ask trainers why they rewarded a substandard behaviour that an animal performed; 'Shame, he was really trying hard.' And all the reward has done is furnish the animal with the incorrect information, which will limit the animal's future potential. It is not kind to be grey when communicating. Clarity is always required. The truth will set us all free.

I worked with a wild horse that had never been touched before. I was doing the initial round pen work where we teach the animal to respond to us, and ultimately follow us around. Our first goal is that the animal moves its hind quarters away from us, and the second goal is that they stand still while we approach them and allow us to touch them. This little animal was blind in one eye. I felt sorry for it and was a little desperate to have it allow us to handle him so that the vet could check out the eye. My pity for this horse affected my judgement. I was making very slow progress and in fact, the horse was looking more and more confused. I left the round pen to consider how to proceed. I decided that I needed to enter the relationship and stop compensating for him being blind on one side. When I did this my progress was immediate. He understood me. My pity was no longer in the way.

With all this in mind, it is important to remember that we can only free ourselves from any detrimental effect the voice in our head might have by firstly recognising it for what it is, and secondly, forgiving it for the judgements it makes. When we are able to achieve this level of humility, we will be able to rise to the occasion of being more present, intuitive, open-minded and objective when training animals.

Chapter 14

Intention – The Universe Handles the Details

Taking care of the details of our lives is generally considered a rational mind activity. However, when our rational minds are active, we shut out our Infinite Intelligence, which has the capacity to handle the details in ways that are vastly more supportive to us and everybody else. (Arnold Patent, Stephen and Kathleen Norval – Universal Principles)

Just Be

While in session, we may become conscious of contraindicated emotions, for example, when feeling frustration we are probably stuck in the 'beta mind state', and not receptive to intuition. We become unfocussed on what is actually occurring, and more focussed on judging ourselves or the animal in light of the proposed outcome. Our empathy and gut instincts have great value as we attempt to understand certain complex behaviour interactions, provided that the behaviour, as it occurs, is recorded precisely and objectively. In light of this, we cannot be sufficiently objective if we are caught up in the beta state, because it is in this unaware judgemental state that we can become victim to our preconceptions. We need to be focussed on the moment, the here and now, so that we can properly intuit what needs doing to move to the next step. I would always encourage trainers that while in session, or observing sessions, to stay present in the moment and with their feelings. If they begin to feel unfocussed, they are directed to switch gears or end the session. This way they can return at another session after they have calmly considered their options.

Clear communication is key to ensure success. We listen, as outlined thus far, using our intuition and observational skills. This is not to say that we must simply go into a session listening and hoping for an esoteric Dr. Doolittle experience. We also need to deliver clear messages to the animal which is only possible if we are clear on what we wish to say. So, before going into a session, be clear on what you wish to achieve. If your plan is to wait for the animal to exhibit behaviour creatively, and then reward that creativity, ensure that you are clear on how you will maintain clear communication with the animal of your expectations. If you are going into a session with the objective of training a seal, for example, to do a back flip in the water, you need to be prepared to complete the training of that behaviour in one session. Chances are it will take longer than one session, but there is always the chance it will take just a few trials before the animal is trained the behaviour.

If you went into the session with no idea on what you wanted, or how you were going to achieve the result, your communication will potentially be hesitant and for the animal, confusing and frustrating. That is why it is advisable to detail behaviour plans. This essentially, is planning

Chapter 14 Intention – The Universe Handles the Details

a training session so that we can visualise the process and most importantly, the end result. This will ensure we go into the session confidently, and this confidence will manifest in our delivery. The easiest way to get into a good habit of this, particularly if the concept is new to you, is to actually write down the steps you will be using to achieve the behaviour. More than anything, this will ensure that one has visualised the end result. Visualising the end result is a very powerful training tool and even an important life skill. Many philosophers and wise people have written that if you can visualise something, you can achieve it. This is because the visualisation of the end product is the beginning of a creative process that ensures you are working towards something.

Visualise End Result

The lady, Wynter Worsthorne that taught the animal communication course that I went on believes that animals and humans think in pictures. So, when we have a thought, the animal picks up the mental image as well as the intention and energy behind the image. This is described in the literature as a 'quantum hologram'. Wynter concludes from this that the more conscious we are of our thoughts and feelings when working with animals, the more effective our communication will be. Wynter elaborates that the animals pick up on our visualisations and that this is the reason why they often do what we don't want them to do. If, for example we don't want the dog to come inside, we more than likely have a picture of the dog inside, and because the dog sees this picture, he thinks we do want him to come inside. After the course, I had an interesting experience. My pig, Cleopatra is a very stubborn animal. She is deaf, and for this reason, I probably overcompensate with her. The easiest way to get her to move is to offer her food, then walk away with the food, and when she follows, to give her the food. Before she went deaf, I used to clicker train her, and she learned some simple tricks. One was to dance. When she was a piglet, she was very agile with this little behaviour, and would twist her body round and round easily until she heard the click. When she became deaf, I stopped training her.

I was late leaving for work one morning, and eager to lock up quickly. Cleo was lying against the gate, and I could not close it to lock it. She weighs over 100 kilograms now, so it was not possible to push her out of the way. I remembered Wynter's theory, and decided to try it out. Rather than visualise what I did not want Cleo to do, I visualised her getting up and moving out of the way. Cleo, it would appear had other ideas on her mind. She was waiting for the food, or so it seemed. But I persisted. I did not speak, just held the image of her moving in my mind. She stood up and looked at me with her gorgeous brown eyes, almost fluttering her long eyelashes. But she would not budge out of the way. I held onto the image, and adored her at the same time. We were having a battle of wills. Then, all of a sudden, she began her dance, swaying her heavy body around and around, and then stopping to have a look at my reaction. She was set on the food. She knew I wanted her to move, but was communicating to me that this was food time. I giggled and went to collect a couple of her favourite cocktail tomatoes. Work would have to wait.

I am not sure if she saw the pictures in my head, but she must have been aware of my focus, and because I usually made a fuss to get her out of the way, my novel presence that day did receive a more communicative response from her. So, who knows? Never say never. The imaginary home that my sister and I used to play in is now a real house where my sister and her family live.

Chapter 14 Intention – The Universe Handles the Details

Nowadays we are all exposed to so much stimulation, and most school systems do not really encourage the development of imagination. Yet, imagination is the way that we are able to visualise the end point. I remember as a child listening to stories on the radio. I would visualise the story as it was delivered. I remember playing out games of life with my sister, where we had the entire home that we were playing in visualised on the back garden. We knew where the lounge, the kitchen and the bedrooms were, plotted out on the lawn. We walked through doors and opened windows, and it was all just in our imagination. Television and computer games have largely replaced these creative visualisation exercises. At schools, our children are being placed on medication to ensure that they remain focussed. This focus prevents the children from day dreaming. Dreaming is a part of the creative visualisation process. As parents and responsible citizens on this planet it is important that we remain conscious of this development in our societies, and that we provide our children and ourselves with opportunities to exercise and develop our imaginations and creative skills. If you can dream it, you can do it.

Visualise and the Universe will Handle the Details

I have a collage in my bedroom that I put together at a point in my life, when I was living in a little complex, my husband was living abroad, I had no pets, and did not own my property. On the collage I had pictures of animals, a feel of country existence, words to the effect that it was time to buy, pictures of families together, and even words of various places I wished to visit overseas. In just three years, ninety percent of the visions on the board were complete. Most of the little pictures I have even forgotten about, but periodically I go back to the board and am amazed at how easy it has all been to achieve, and the achievement has simply been because I visualised it.

To have this book published is a part of that vision. In fact, some of the manifestations have been with so little effort that it is remarkable. I mentioned to my husband after he returned to live with us in South Africa that we should buy a house. He mentioned it to a mutual friend of ours an hour later, by chance. It turned out that she was selling houses at the time, a fact that neither of us knew. She called me an hour later and said she had a small holding in the country she wanted to show us. We signed papers to purchase the home that evening. No effort required. The imaginary home that my sister and I used to play in is now a real house where my sister and her family live.

When working with animals, the same applies. If we can visualise the end, and some of the process required to achieve the end, we will communicate effectively in order to achieve that end. On the other hand, to go into a session without an idea of what we are going to be doing has the potential to result in confusion for the animal because we will appear hesitant, our thoughts will be scattered, and we will be more inclined to chop and change our mind.

If a session is not going as we imagined, it is however important to remain flexible. However, this flexibility, in our experience, is easier to apply if we have that plan. Because we are prepared, and more relaxed going into the session, we are more easily able to intuit what is occurring, and adjustment is thus easy. It is the SLAM principle in conjunction with the visualisation. Remember, the stop, look/listen, adjust and move principle. We know where we are going, we are receiving messages from our scenarios that are helping us to see where we are on the road to where we are going, and these messages are telling us to stop, see what is going on, and then make a plan, and implement that plan so that we can get back on that road to where we are going again. There must be no time line or effort that you are striving for, because this is working against the situation. We must

Chapter 14 Intention – The Universe Handles the Details

remain flexible in order to recognise when adjustments are required. If we remain fixed on the end point, we enter into power play.

This is the difference between goals and intentions, and it is important to make the distinction. An intention is something that we know is going to happen, while a goal is something we strive to make happen. When we consider the power struggle that is the trap that so many of us fall into when training animals, this is a result of us fixating on the goal, and putting a time limit on achieving the goal. It seems that the harder we try, the less we look as though we are achieving. That is because we are focussing on how we think it should be rather than seeing the situation for what it is and adjusting to that as is required. If we simply intend behaviour, we are confident of the result. We will adjust where required because we will not be tied into the striving. We will not be focussing on trying because of the confidence we have in the outcome. The important thing to note in this is that the focus will essentially be on the relationship we have with the animal, rather than on the achievement of the behaviour goal. It is also the difference between being subjective and objective. When we become subjective and goal driven, we take things personally. When we simply and confidently intend a result, we don't need to take anything that happens personally. It is just a by product of a wonderful relationship that we share with the animals.

Chapter 15

What we Focus on Expands

Energy flows where attention goes. Some refer to this as the universal law of attraction.

I am drawn to puzzling situations with animals. These usually result in trainers calling the animals involved troublesome. These are the animals that teach us the most. They are usually not troublesome at all, simply very bright. I have alluded to Affrika and Zulu earlier in this book. The following example is once again about this duo. I had not worked the pair for a while. One of the senior trainers asked me to watch a session as all the trainers were beginning to experience problems with this pair of dolphins. The animals were swimming off in the sessions and taking turns to co-operate. The trainers were confused about how to respond. Five different trainers were working with the animals, and to varying extents, all five were experiencing the same problems. One of the trainers expressed her feelings, 'I think that the dolphins are bored. I find that if I make the sessions more exciting by running around faster,

Chapter 15 What we Focus on Expands

and trying new and different scenarios I get a slightly better response.' A second trainer retorted to this remark, 'I am not sure about that. I found that when I did that it worked a week or so ago, but it has not worked the last few times I tried.' These comments set off warning bells in my mind. It sounded as though the trainers had begun compensating for the dolphins rather than addressing the route cause of the problem. My concerns were well-founded. When I stood in front of the dolphins, I noted that they appeared unsettled, and even a little anxious. I watched a few training sessions.

The first session I saw I noted that Affrika kept swimming off. The trainer watched Affrika go, and gently held Zulu's rostrum in her hand, waiting for Affrika to return before she continued to do the session. It was clear to me that Affrika was receiving as much attention as Zulu was with the trainer's eye contact firmly on the wayward dolphin. When I questioned the trainer, she said that when she gave Zulu attention if Affrika was drifting, the session became even more difficult to manage. The trainer felt that at these junctures Zulu was actually trying to keep Affrika away. In another session I watched both dolphins begin to lose focus. The trainer immediately picked up the cooler boxes which were filled with the fish treats she was using in the session and put them in plain sight of the dolphins. Before she did this, the fish had been out of sight behind the wall next to the pool. Both dolphins temporarily provided the trainer with their attention, which she correctly rewarded them for. What the trainer failed to notice in this play of events was that the dolphins were actually rewarded for their intention to swim away by her action of lifting the cooler boxes into view.

In all the sessions that the senior trainers and I watched that day, it was clear that in their attempts to ensure the cooperation of the dolphins, the trainers also appeared flustered and anxious. This scenario had been playing out for a month or so, and according to the trainers, seemed to be getting worse by the day. It was clear to me that the foundations of the relationships between animals and trainers were shaky. Because these two dolphins took part in daily shows, the trainers were under pressure to achieve their co-operation for the sake of the quality of the shows. This made the anxiety the trainers were suffering even more real. As much as we would like to think that we are not affected by the opinion of an audience, it requires a conscious effort to retain presence of mind when you are standing in front of over a thousand people and the dolphins refuse to do anything for you. It is a very well developed character who sees this for what it is and does not take it personally. Therein lies a lesson all on its own.

Our Misguided Focus Directing their Action

After further discussion with the trainers, a conclusion became clear with regard to the problem we were experiencing with Affrika and Zulu. We recognised that the dolphins were mirroring the anxiety of the trainers. Furthermore, the dolphins were actually confused. If one of them was participating in the session and the second was swimming around, the trainers would very often focus their attention on the one that was swimming around while they waited for it to return, and not give sufficient attention to the one that was in front of them. Their attention was on the drifting dolphin and so that dolphin was essentially being rewarded while the one close to them was not being offered the same measure of attention.

The dolphins don't necessarily respond only to positive attention. We all respond to any type of energy. Any parent will recount stories of how their children always seem to play up when the parent is tired, perhaps after a day of work when all the parent really wants is to sit quietly and focus on

Chapter 15 What we Focus on Expands

nothing in particular. That's when the kids will misbehave. And the parent will often respond by letting the child know in no uncertain terms that their behaviour is unacceptable, but usually the parent's reactions are not effective. Because they are tired they will probably be more about big talk than acting. The child will not be enjoying the deliberating, however, at that point. For the child, any attention is better than no attention. The child just wants to be acknowledged. It does not worry about the manner of acknowledgement. Negative or positive, it is still attention.

Another problem that was happening with Affrika and Zulu was that every time the dolphins looked as though they were going to swim off, the trainers would quickly try and do something to ensure that they did not swim off. They would ask them to participate in behaviours that they already knew, such as stationing their rostrums on the side of the pool, or on the trainer's hands. Another way to try and keep their focus was to try and move to another area or lift their cooler boxes, or ask them to do an exciting jumping behaviour or similar. All these efforts were confusing the dolphins. They were no longer clear on where they needed to be. They were actually being rewarded by the trainers every time they looked hesitant and uncooperative. That would be the same as the child being threatened with potential punishment, but that punishment not being carried out. The child is receiving mixed messages, and is in fact being rewarded with attention for misbehaving, with empty threats. The attention is there.

When to redirect...

At this juncture it is important to note that redirection is a tool that can be used to prevent an animal behaving in a manner we don't want. By redirection we mean that if we anticipate that the animal is about to do something we don't want it to do, we can ask them to do something that is inconsistent with the unwanted behaviour. For example, as outlined above, when Affrika and Zulu were about to swim off, the trainer asked them to station, which is a behaviour that is inconsistent with swimming off. If, however, the animals keep doing this and the situation is dealt with over and over, then we need to recognise that the swimming off, or the intention to swim off is being rewarded.

The classic human example is when we watch an eighteen month old child. Parents are always very concerned about their children swallowing little stones and other foreign objects – understandably so. But at this young age, the way a child investigates its environment is to place everything in its mouth. This usually results in a parent getting up and taking the object away and trying to redirect the child to an activity that is not related to putting anything novel in its mouth. I have seen so many children of this age, including one of my sons, manipulating their parents with a technique that they have unwittingly been taught by their parents. The picture is usually the parent in company or on the telephone or participating in an activity where it is not 100% focussed on the child. The child will find the target object, look at the parent to ensure that they are watching, and then the child will smile as it places the object in its mouth. The parent will stop what they were doing, jump up, take the object away, and reward the child with attention – number one, and usually another game – number two. The redirection is no longer working. It is now a reward and not a redirection.

Chapter 15 What we Focus on Expands

Reward only what you want

In the case of Zulu and Affrika, we had to simplify their world for them once again. Go back to basics and re-establish our boundaries. The first step in rectifying the problem was to take the dolphins out of shows. The amount of trainers working them was cut back to two. In the first two days, all we did as trainers was concentrate our attention and energy on the space directly in front of us. If they dolphins came into our line of sight, we rewarded them. What you focus on expands. This focus was in the literal and energetic, or calorific aspect. Literal in that we were focussing with our attention, our eye contact and calorific in that we were rewarding their attention with fish. They quickly understood this simple concept.

Our next step was to ask them to participate by doing behaviours that they knew, and then when they returned into the focus area, they would once again be rewarded by our attention and fish. They were then asked to stay in the focus area for variable lengths of time before they were reinforced, either by self-reinforcing high energy behaviours, touching and stroking, or fish. They would never be reinforced with any of these things unless they were in the focus area. This took a day or two for them to understand. The only action was to focus on where we wished them to be. Again, what you focus on expands. We had been making the mistake of focussing on them drifting off rather than focussing on them participating the way they should. They were always only doing what we asked them to do.

When other trainers were introduced into the scenario their performance in front of the animals was carefully critiqued. It was interesting to find that if the trainers became anxious, the dolphins did not cooperate as well as they did when the trainers were calm. The anxiety of the trainers was so clear if we looked closely; a little sloped shoulder in their body movement. In some cases, trainers had to actually break bad habits that they had developed during the last month or so. Some reported that they had to literally and very consciously hold their hands at their sides, as they were tempted to lift them to call to the animals with cues when they felt they were going to drift off. This is such a classic tale of 'who is training who.' The dolphins had taught the trainers to respond by lifting the cooler boxes or running madly around the pool and other subtler body movements. If the trainers anticipated a drift off by an animal, their body language demonstrated this concern. In this scenario, a classic tale and good example of most scenarios, nothing had to be changed about the animal. The problems had developed in the trainers and they had to change their behaviour, attitude and focus in order to rectify the problem.

Entrenching Behaviour

Focussing on a problem rather than doing what needs doing to rectify a problem is a common mistake. I have met trainers who will ask me to watch a problem being played out between themselves and an animal. Language in these instances is along the lines of 'look what he always does when …' This is followed by the demonstration of that unwanted scenario. I have seen those same trainers demonstrating that same problem to more than one person. In these demonstrations, they are effectively rehearsing the problem; entrenching it in the animal's behavioural repertoire. This is a serious mistake. So serious, that in some world renowned facilities, for example, if a marine mammal such as a sea lion, dolphin or whale shows aggression twice and in some instances only once, that animal will no longer be asked to participate in the behaviour where that aggression can occur again. In one facility that I visited, there was a section for the sea lions where the trainers would only work with the animals from behind a fence in what they call a protected contact situation. Sea lions

Chapter 15 What we Focus on Expands

were put in this area if they had bitten their trainers. The philosophy behind these decisions is that the aggression has been rehearsed. Similarly, a predominant school of thought in the pet dog industry is that if a dog bites a human it has to be put to sleep. So, aggression aside, if we look at any unwanted behaviour, if we allow that problem to be rehearsed, we will make it worse.

If something is not working, we need to change what we are doing. If we keep doing what we are doing, we will keep experiencing what we are experiencing. The problem is not with the animal. The problem is that we are focussing on the problem. Look at the problem and determine what needs doing. Try something new. When working with a seal called Boom Boom, I had a problem asking her to leave through a gate at the end of her training session to the enclosure that she shared with eight other seals. She would growl loudly, and if I did not have a barrier to protect myself, she could easily have bitten me. The problem was becoming apparent with five of the other trainers, and we were very concerned that it was going to result in her imparting a nasty bite. There is a medical condition attached to seal bites that warrants mention at this point. It is called 'seal finger'. Men who were bitten while clubbing seals used to often lose fingers as a result of these bites. If people are bitten by seals, medical opinion dictates that they need to endure a harsh course of tetracycline antibiotics to prevent the condition.

Extinguish unwanted behaviour

Boom Boom was normally a very calm and co-operative animal and we did not want her to begin biting us. She was an excellent show animal, and very easy to work with in all scenarios. We decided to try and change the focus. Almost reverse psychology. The growling noise became our focus. Just before gating her back we asked for the growl by issuing the verbal cue, 'grumble'. When she delivered it, at first as an aggressive warning, we bridged and rewarded the behaviour. Soon she saw that this had become a game that would result in her being rewarded. When we had the behaviour on cue, we stopped asking for it at the gate, and very soon she was singing into a microphone during the show, much to the amusement of the public. She no longer growled when we asked her to gate back, and this would also be intermittently rewarded once she gated back nicely. This training technique, theoretically is termed extinguishing a behaviour. Energetically, it is simply focussing on an outcome that we wish to achieve rather than putting all our efforts into avoiding a situation.

My husband, two sons and I have five dogs at home. We live on a plot outside town, which is lovely except for a nasty driveway. I am very concerned about the driveway as it is very steep, and once you have committed to driving in or out of the gate, there is literally no turning back. With dogs in the way, there is the possibility of injuring them, something we are not prepared to risk. So, the dogs are locked away if we are driving in or out. This resulted at one point, in the gate being a very exciting place to be. Two of the dogs were intent on getting out of the property and rushing off down the road. The problem seemed to be getting out of control until I saw the reason why.

Every time the dogs got out my boys and husband would do a frantic screaming charging rampage down the road. An objective dog god in the sky would have seen a fantastic game between dogs and dog people. On one occasion, when I did just let one of the dogs go without the colourful entourage, he was picked up by the neighbour, at least a kilometre away, and given a lift home. He hopped out of the car after his adventure, filled with the proud joys of spring. Very soon we had a problem with four of the five dogs instead of just two. The dogs were having a fabulous game at the gate. My husband loved to quip that they were a typical animal trainer's

Chapter 15 What we Focus on Expands

dogs, completely badly trained. A lecture on 'what you focus on expands' had to be carefully deployed to the family so that the problem could begin to be rectified. It would not work to focus all our energy chasing them down the road. That game had to be reserved for inside the gate, where we wished them to be. Over quite some time the behaviour of playing outside the gates was replaced by a game inside the gates if they did not make the escape to the great beyond.

Chapter 16

Ego – Friend or Foe

Does fear prevent you from living? Or does living assist you to embrace your fears? It is all determined by how we are in relationship with our egos.

The clarity and success of applied behaviour modification principles is coloured by the human personalities who apply them. These principles are our communication system with the animals. Communication commonly includes talking, listening and body language. Talking is the use of cues, bridges and other requests for an animal's cooperation. Listening is sensing the temperament of the animals which enables us to set them up to succeed. As we have learned, effective body language is fundamental to successful animal training because animals have an astute life preserving natural ability to read posture. Humans unconsciously express their emotional states in their posture. Consciousness of how we are feeling is thus vital to ensure that we are not communicating mixed messages. For example, new trainers may come across

Chapter 16 Ego – Friend or Foe

hesitantly because of their lack of experience and confidence resulting in animals not responding appropriately. The animals respond slowly, and appear to be 'messing the trainers around'. To be consciously 'self-aware' enables us to note our effect on the animals and ensure productivity.

Being in a fulfilling relationship with an animal is the foundation. If this is not the case, then why bother. Not only will it be fulfilling, if we consciously keep it that way, but our training progress will also be more productive. Technically relationship is generated through appropriate animal training practice that results in the trainer achieving 'secondary reinforcer' status. "Control" behaviours are built onto these foundations. These include anything that assists us to train an animal to do something else. For example, teaching a dog to sit and stay would be foundation control behaviour. Other control behaviours would include teaching a rhino to target, or with the dolphins, teaching them to station, focus or gate from one pool to the next. With horses, control behaviours would include having them allow us to physically manipulate them into various positions. If the basics are in place, it is easy to train what can be referred to as the decorations – any other behaviour. So, for that dog it may be fetch a slipper. The rhino may be taught to stand for a blood draw; the dolphin taught to jump out of the water a series of times. The horse will be taught to draw a carriage.

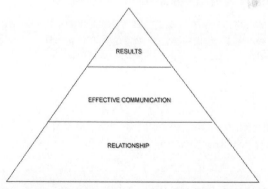

Relationship breaks down when we are not focused and present.

What is being Present all about?

Being present means focusing on the task at hand without falling victim to any productivity limiting distractions. Besides being necessarily focused, being present is also important because in this feeling and connected state we are able to be intuitive, which is required when gauging the animal in order to set it up to succeed. The beta state is a rational state that occupies our minds when we are analyzing and judging. Being present means that we are in the moment. When we are in the moment, we see 'what is' as opposed to 'what we think is'. 'What is' refers to the actual events before us, and not our judgments of what is before us.

What is it that causes us to lose the moment?

The ego. When faced with a challenge the human ego responds in one of three ways.

Power Play – Digging in our heels and insisting that the animal responds on our terms.

Same Old, Same Old – Doing the same thing over and over and expecting a different result.

Throwing in the Towel – Giving up, escaping from the challenge, usually with an excuse.

For example, say we are training a seal to lie still in a position where a vet will eventually insert a needle into the vein in his hind flipper in order to achieve a blood sample. The seal however keeps fidgeting in the position, which is a problem, because when there are needles around we need the animal to keep still, for its own safety. If we simply work with 'what is' we adjust our training to help the animal succeed. The ego involvement on the other hand, could result in the following:

Power Play: Restrain the seal and force the issue

Same Old, Same Old: Do not change anything and expect the seal to do the right thing

Chapter 16 Ego – Friend or Foe

Throwing in the Towel: Give up, usually with an excuse as to why the seal cannot do this behaviour. I have heard the most elaborate and creative excuses in instances such as this, including: He does not like the vet and he has very ticklish flippers.

If we look at this closely, we will notice that we don't only respond in this manner when relating to animals. It is the classic human condition to respond in these ways.

When animals face challenges, they respond with fight or flight, simply 'expressing' themselves in the moment; handling 'what is' and being present. We concern ourselves with 'impressing' others. To be more effective trainers we would do well to recognize that at the challenge junctures, we have a choice – we can go one of the three ways documented above, or, as in the case of the animals, we can stay present, deal with 'what is' and do what needs doing. So, let's take a lesson from the animals and see how to respond in the moment – expressing our great intelligence rather than trying to impress, which is all about the future.

	Impress - Distracted	**Express - Present**
How we respond to animal's doing something inappropriate	By being right - power play, same old or giving up.	By being true - responding to 'what is'
What we focus on	What is wrong	'What is'
Our responses are	Rigid	Flexible
The effort is based in	Thinking (beta) - not present thus poor communication	Feeling (alpha) - Able to respond in the moment. Able to be intuitive.
The goal of our action is	Excuses	Solutions
Our ability to take risks	Unwilling	Willing
Outcome	Insecure, defensive, aggressive, over confident or unconfident trainer	Confident, objective and able trainer

Touching Animal Souls

Let's define 'ego' as the creator of stories, which are more commonly known as distractions and excuses. These distract us away from being present and leave us less able to participate as effective trainers. The stories originate in one of four places; the past, the present, our feelings about ourselves, or our feelings about the animal.

Past: These are past experiences and memories which are experienced as regret, guilt or nostalgia. Being aware of the animal's reinforcement history, without letting the emotion of the information affect our ability to be present is what is required. I may be afraid of a dolphin because he once bit me, so I am nervous, expecting the worst. To succeed with the animal I need to deal with my fear and consolidate the relationship so that the animal never suffers my confusing hesitance. I need to work within my limits, to secure my confidence and presence of mind before progressing.

Future: This is in the shape of fear or hope. Hope could lead us to get ahead of ourselves in the successive approximation process. If I am over eager for an animal to succeed, I am not intuitively clear that he understands, so may confuse him by omitting steps. Alternatively, fear of an outcome may hinder my ability to sense that the animal is ready to move to the next step. I worked with a rhino keeper at the Giza Zoo. The keeper was an incredible man, and dedicated to his charge. He and the vets in charge of his section were concerned that the female rhino who had been at the zoo for over twenty years would not go into the night quarters without a dangerous fuss, and this meant that the keepers had to clean the area with the rhino on exhibit, which in itself was dangerous. There was a high wall surrounding the area which the elderly men would have to scale when jumping in and out of the camp. This made their escape difficult should the animal decide to charge at them. I watched carefully as the men went about their normal day of cleaning. In order to do the mucking out in the safest way possible, they would put food

Chapter 16 Ego – Friend or Foe

on one side of the area and then proceed to clean on the other side. When this part of the exercise was finished, they would duplicate the process after putting food at the other side. They told me that the rhino would not even eat from their hand. They described her as skittish. I was in attendance just before they were going to begin to clean the second part of the exhibit.

The rhino was standing in the exhibit and I was watching her as the keeper, Mr. Fathay walked around the area with a big bundle of grazing for the rhino. As he went he called out to her from time to time. Whenever he did this, I noticed that the rhino's ears were alert and following him around the space. When he reached his destination and was about to toss the food over the wall, I heard him change his tone, and so did the rhino. For the first time she lifted her head and began to turn it in his direction. He tossed the food over and she trotted off towards him to eat her second course. For me the training plan was instantly in my head. All Mr. Fathay had to do at the next feeding was to wait until she turned and took one step towards him before tossing the food into the exhibit. He was to repeat this every time waiting for her to progress a little closer before the food was rewarded. Eventually the animal would be close enough to eat from his hand, and then they could begin target training. Ask the rhino to touch a target pole, use a bridge and then reward her by hand. This exercise could be used to teach her to enter the night quarters. I imagined the training would take a month or so and they would soon have their wish. Mr Fathay however had a different time line in his head. He sat at that exhibit the entire day and that very afternoon, for the first time, the animal was eating out of his hand. His enthusiasm for the project was truly inspirational.

The next morning they radioed me from across the zoo to come take a look at the rhino in the night quarters. I was concerned. The rhino was too. The keepers had locked her in the first time she ever entered the space for them. They were about to embark on a large cleaning spree. I asked them to please release the rhino directly. I feared that the trust that Mr. Fathay had generated the day before may be broken. They had not taught the rhino to accept the night quarters one step at a time. They had trapped her there the first time in order to get their cleaning done. My fears were well founded. The rhino's training suffered a setback, and it took her a while to progress back to eating out of the keeper's hand. The good news is that this lesson was one well learned, and the keepers have managed to get the relationship back on track after a month of going back to basics. The rhino is now in the night quarters willingly.

Subject Consciousness

Personal feelings of inadequacy take us away from being present. Over confidence, giving up, being a victim, blaming others are a few of the results of our personal reactions. Many of us see ego as a large blustering personality that is dogmatic and forceful. However, ego can also take the shape of someone meek and hesitant. Someone who believes they are less than and not worthy. This will all come out when the trainer stands before an animal. This person may train animals with a hidden motivation to achieve approval from their peers. Thus they are more focussed on a colleague's approval than on the animal, which will result in a lack of intuitive insight, affecting timing, cue clarity, and much more, so reducing potential to achieve training progress. The big bolshy ego will also have an idea in their head about impressing others, but this would be played out differently. The effect would be the same lack as in the previous case. It will probably have a different result, but in essence, still be ineffective in the long term. These may be the characters that will choose punishment, the my way or the highway approach as opposed to the meek trainers who will choose the giving up or same old, same old methods of dealing with a challenge.

Object Consciousness

These are anthropomorphic concerns that affect our ability to succeed. If, for example I am concerned about hurting a dolphin in a blood taking procedure, my tension will translate while training the behaviour, stifling progress or developing excuses not to succeed.

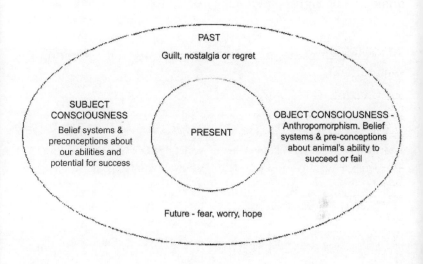

Leadership vs Dominance – Which Style is Self-Aware?

As we have discovered, the difference between being a leader and a dominator is more subtle than our choice of reinforcement over punishment. We cross over from being a leader to being a dominator when our egos overtake our consciousness. For example, a dolphin continually provides us with the incorrect response, a breach to a spin cue. We continue asking in the same manner making no adjustments. Our dominator excuse – "he is naughty, he knows what I want". If we continue this practice we sour our relationship with the animal because communication is not clear. A leader inspires the outcome through guidance, senses whether the animal is confused and adjusts accordingly operating in the moment thus expressing themselves. A dominator as a result of their inherent need to impress, potentially based on deep seated feelings of inadequacy, is reactive (insisting or leaving) as opposed to active (responding to 'what is'). A dominator has a need to be right, is defensive or aggressive, inflexible and unable to focus in the moment. A dominator cannot do the simple fundamental activity of training – ignoring the bad and rewarding the good because they are not operating in the moment.

LEADER – EXPRESSING	DOMINATOR - IMPRESSING
ALTRUISTIC - for relationship	**NARCISSISTIC** - looking good
PRESENT - focussed on task at hand	**ABSENT** - concerned about what others think
FEELING - focussed on task	**THINKING** - past, future, subject, object
SUFFICIENT – trusting and self aware	**WANTING** – untrusting and afraid
FREE - flexible within the goal - about relationship	**ENSLAVED** - beholden to the goal
TRUE – authentic self, clear	**PRETENDING** – victim or abuser, masked
HONEST - share and open to feedback	**FALSE** – defensive, making excuses
INSPIRES – provides choice and guidance	**DEMANDS** – insists on their terms

Chapter 16 Ego – Friend or Foe

Inter-Human Ego

Many of us train in groups of trainers. Where there is more than one person operating with the animals, there is a greater risk of ego getting in the way. An effective functional team is necessary to get the training done. It is my experience that a dysfunctional team has a collective ego which can be even more detrimental to the animals in their care. It results in disharmony in the work place, and guarantees lower productivity and even regression with animal training. To truly put animals well-being first is only possible if we all humbly put our egos aside. Animal trainer teams are usually also responsible for more than just training. At the marine mammal facility trainer's responsibilities include cleaning, fish preparation and stock control, water chemistry, record keeping and show presentation. This is similar to many zoo and domestic animal settings. This means that it is imperative that trainers support each other so that we can be present when we stand before an animal.

If we have pets at home, our duty to them is no less. Homes with disturbed personal relationships between the humans will rub off on the animals. I watch my children excitedly play with their friends in the garden, and the dogs are a bundle of energy around them. The daschund, Jack is the commentator, and barks with abandon. The other four dogs bound around the kids, play fighting with each other, just like the children are doing. When my husband and I have been in arguments, I am always aware of a sombre looking lot of dogs lurking solemnly in the background. In fact, there have even been many case studies where domestic animals have taken on the physical symptoms and complaints of their humans.

What can we do to Overcome and Diminish Ego?

Responsibility and Accountability - As we have discussed, the difference between being a leader and a dominator is ego related. Leaders provide responsibility which is providing their charges with the ability to respond. Recognizing in ourselves if we are providing choice while training an animal is the difference between us being a leader (present and accountable), or a dominator (absent and demanding). Responsibility also has implications in the way we problem solve. As trainers who wish to be effective, while problem solving, we create opportunities where we have the ability to respond. We focus on what we can do as opposed to what we cannot do – the excuses. We then empower ourselves to be accountable for our actions and the reactions of the animals.

Tell the Truth about what we do. If we don't tell the truth about what we do, we cannot be clear in our use of the methodology. Misunderstandings in the general animal training theory create these false conceptions and confusing protocols, which, in turn, result in less progress. Let us look at three misconceptions that have crept in over the years. They relate to the leader dominator concern, negative reinforcement and the concept of ending a session on a positive note.

Leaders: The historical understanding of an animal trainer being a chair wielding, whip cracking dominator has left many trainers concerned about the animals being relegated to a submissive role in our relationships. We hold the resources so we are in charge. However, how we choose to use those resources determines how we are in relation to the animal. When we are clear that we are leaders and not dominators, the submissive status no longer has any bearing. A respectful relationship based on sound communication has been generated.

Chapter 16 Ego – Friend or Foe

Negative reinforcement: The definition of this and other relevant terms is detailed in the appendix of this book. Suffice is to say that most trainers I have come across do not understand the concept. There is a general misunderstanding about what negative reinforcement is all about. Many confuse it with punishment and so have shied away from using the tool. Historically the definition stated that the stimulus that would be removed when the animal responded according to our needs needed to be aversive; meaning this required an anthropomorphic need to be 'nasty'. More recently the definitions are being amended to exclude the word aversive. We are now able to apply the concept without feeling 'guilty'. Whether we know it or not, most of us do use negative reinforcement even in scenarios where we primarily use positive reinforcement. For example we train the dolphins to lie stationary, in a prone position, with their tails on our laps.

We train the dolphins to lie in this position so we can take blood samples from their tail areas. Most of the rest of their bodies are covered with blubber, so blood vessels are difficult to find. The tail is used by the dolphin and other cetaceans as a thermal regulation device, so the veins are close to the surface of the skin. I always find it amazing that we are able to teach animals to do this type of behaviour. Remember that a dolphin's tail is its engine, so a very powerful part of its anatomy, and we ask the animals to place it on our laps during this procedure. If we are clever with the training, and work on the trust between ourselves and the animal, we achieve the desired result. I have seen a few dolphins make no bones about the fact that the training was not successfully achieved, by flicking out of the position and injuring trainers in the process with those strong tails, or even by retaliating with a bite. However, when the training is successful, the animals are incredibly cooperative. It does serve us and them if they are, as we are more easily able to achieve their care if we proactively take these samples. We can monitor their health when they are well and achieve normal blood results. When they are not well, we have a reliable reference to compare their blood results. On more than one occasion I have been completely bowled over by the fact that a dolphin has cooperated with us to provide this behaviour, even though they have not been well, and even on occasions, when they have not even been eating. Nothing replaces relationship. This is a very vulnerable position for a dolphin. It is reinforced for being calm when we let its tail off our lap after a blood position, as it is restored to its natural state of being.

Understanding that we use negative reinforcement means that we are more inclined to use the natural behaviour of the animal as part of the reward offered which further requires us to be more conscious of the natural inclinations of the animals we train. When training horses, negative reinforcement may actually be a primary reinforcer. Primary reinforcement is something that is naturally or biologically rewarding. For humans, this would include food, shelter and sex. For many animals, the same applies. Using negative reinforcement with horses, may in some instances, particularly when doing ground work, be the formulation of a relationship where we provide clear leadership to the animal, thus providing it with a sense of security. This sense of security is something they would naturally achieve in relationship with a herd where they negatively reinforce each other for various actions. When a more dominant horse wants to feed in a place where an underling is eating, he simply moves in that direction. The less dominant animal will give way willingly.

Presumably the weaker animal has then aligned himself with the more dominant animal, thus affording him some security in the herd situation. At the Prague zoo, there was a Kiang, which is a wild donkey. Franta, the trainer was called to work with the animal because it kept charging the fence when the public came near to the exhibit. This could have been dangerous to the donkey. Franta had to use negative

Chapter 16 Ego – Friend or Foe

reinforcement to teach the donkey to allow him to get close to the fence. He would retreat from the fence every time the donkey was calm. I have a wonderful video of him cycling through the zoo up to the fence; something he did this a few times a day. Eventually the donkey was calm enough for Franta to begin providing positive reinforcement.

End on a positive note: Ending a training session or interaction with an animal on a positive note is something that the literature tells us is a good idea. It makes sense as then the animal will remember us as something rewarding. We need to ensure that this does not limit the spontaneity of the interactions, as the result could be dull. As an aside, a point to remember is that if we are able to leave a training session at any point, our ego is not involved in the session. This is a valuable question to ask ourselves while in a training session.

Don't get Lost in the Bridge – Go with the Flow

Animal training is about enjoying the journey and not only the destination. To only be focused on the end goal not only takes the enjoyment out of the task, but can remove us from the now. Our training goals may be important but we must never forget to listen, sensing where the animal is at and setting it up to succeed in every moment. When in the moment, we are more flexible which usually enables us to achieve more than we anticipated possible. I learned operant conditioning using positive reinforcement at the beginning of my training career. I was never taught anything about posture or attitude. When I began training horses, I was introduced to an art form of training; being conscious of my body posture. In my yoga classes I have been coached to be conscious of my posture. This body consciousness has done a huge amount to improve my abilities. Until

I was introduced to the world of horse training, the magic was only in my bridge. I would wait for the opportunity to blow the whistle or click the clicker or tell the animal good, without for any moment considering how I was potentially affecting the animal's behaviour through my attitude and posturing. This is effectively becoming lost in the bridge. For seeing the relationship in terms of what we can get out of it rather than how we can positively affect it through a greater level of consciousness.

Don't take their natural inclinations personally

When we are more present in relationship with animals, we begin to note what is naturally motivating for them. This varies from animal to animal. Gandlaph, my horse friend at home enjoys a game of sniffing new things. Mushatu prefers gentle massaging stroking. If I tried gentle stroking on Gandalph, he would try and entertain me with a game of mutual grooming. We can use the idiosyncrasies of individuals as well as more species specific motivators to reward animals. Letting a high energy dolphin give expression to this energy by sending it to do a large jump is a potential reward. Noting their natural inclinations can also help us to understand what is going wrong. We experienced difficulty trying to train the dolphins to bring their toys back to us. On a number of occasions, trainers tried to snatch the toys away which resulted in the dolphins launching for the toys with enthusiasm. This is probably because of their natural inclination to catch fish. Horses in general often like to stand around feeling secure and grazing. I was asked a question at a stable yard about a horse called Ringo that was being treated for a leg injury. Ringo is trained for endurance, and a star at what he does. This horse was being kept in a very small area so that his leg would be well rested during his recuperation phase. When the animal had to be led from one area to another, it

Chapter 16 Ego – Friend or Foe

was understandable that this forward moving horse found his freedom from his space exhilarating, and he would prance excitedly and in effect take the person leading him for a walk. Thus, this premise of the horse wanting to stand still and do nothing was not the case with Ringo. What could the trainers do to teach the horse to walk under control? They could not put any undue pressure on the animal because of his injury. They basically need to look at what was motivating the horse – moving forward. This was their currency. The plan that was initiated was as follows. When leading Ringo, if he walked calmly, the walk would continue. If he pranced ahead of his trainer, the trainer would stop. Eventually Ringo worked out that the only way to keep walking was to walk safely and respectfully behind his trainer. The trainer never yelled. Just the trainer simply standing still and only walking on when Ringo was calm, was enough to get through to this superstar. I have seen so many horse trainers do the opposite. Yelling and chasing their horses when this type of situation occurs. Ego, fear and taking the animal's natural inclination personally.

Horses, we generally teach, like to stand around feeling secure and grazing. I was asked a question at a stable yard about a horse called Ringo that was being treated for a leg injury. Ringo is trained for endurance, and a star at what he does. This horse was being kept in a very small area so that his leg would be well rested during his recuperation phase. When the animal had to be led from one area to another, it is understandable that this forward moving horse found the freedom from his space exhilarating, and he would prance excitedly and in effect take the person leading him for a walk. Thus, this premise of the horse wanting to stand still and do nothing was not the case with Ringo. What could the trainers do to teach the horse to walk under control? They could not put any undue pressure on the animal because of his injury. They basically need to look at what was motivating the horse – moving forward. This was their currency. The plan that was initiated was as follows. When leading Ringo, if he walked calmly, the walk would continue. If he pranced ahead of his trainer, the trainer would stop. Eventually Ringo worked out that the only way to keep walking was to walk safely and respectfully behind his trainer.

The SLAM principle assists me to keep my ego in check. This concept has been introduced already in this book. It is a key concept which enables us to stay focused on 'what is'. The key point in the SLAM principle is being accountable, acting to solve the problem, as opposed to becoming a victim to what we cannot control. Monitoring if we are succeeding and going back to the drawing board where required keeps us moving towards a solution rather than frozen in failure.

Animal Training is a Lesson in Self Awareness

I have shared many life lessons in this book. I am in the fortunate profession where I am more successful if I am self-aware. To sum up, I would like to briefly outline how this occurs in every moment I spend in relationship with animals. Here are some examples of how they keep us conscious.

Mirror Principle: This book includes a chapter on the mirror principle. To sum up, it can be said that when we relate to human beings the feedback we receive is from someone who has their own agenda. It can be likened to standing in front of a mirror that warps our reflection. When in front of animals the feedback or reflection is clear. If we are not clear and present, they will let us know, because our training will fail. The reason we are not present is for us to determine, and if we are willing to look at the reason, we will usually note that it takes us one step closer to being a more conscious human. For example, if I am prepared to look at why I feel threatened by

Chapter 16 Ego – Friend or Foe

my peer's opinion of my training, I will be able to do something about my concern. This will make me a better human, and a better trainer.

It is Just Feedback: Animals are free of agendas, even though most of our excuses discount this. Their success or failure or, vices and virtues in relationship with us are always just an indication of whether we have succeeded in communicating effectively. We can take this personally or use it as information on where adjustments are required. The same is true when relating to people. The more open we are to feedback, the less inclined we are to be defensive. If the feedback is inappropriate, we don't need to integrate it. The SLAM principle, when it is a way of life, creates a life that works. As animal trainers we get to practice this way of life daily.

Ego Triggers: We may not be aware that we even have a concern until we begin to notice where in our training sessions we tend to make the most excuses. At this juncture, it is valuable to begin to look at what belief system or preconception we are using to limit our progress. For example, if we find that we do really well when training behaviours that take a couple of sessions for the animal to understand, but are challenged when the animal does not pick up on the concept quickly, we may need to investigate if this pattern occurs in any other part of our lives. Do we battle with consistency when the going gets tough? If so, why? If we consciously work through this by disciplining ourselves, that pattern in our lives will shift. This has been our experience in the mammal training team at Sea World. As trainers have become more accomplished and experienced, their general work ethic and application has improved with animal training and in other aspects of their work.

What we Focus on Expands: When using reinforcement we continually focus on and reward the desired behaviour. We ignore what we don't want. Imagine a life that works in the same way – where our glass will always be half full or even over flowing, because we will no longer notice or attend to anything less than that. As animal trainers we get to practice this for a living. Practice makes perfect.

We can only Control Ourselves: When in a 'present-focused' relationship with animals we begin to be conscious of our moods and feelings. Ultimately we are able to choose our attitude resulting in greater measures of happiness and success with animal training and in our lives.

LRS – This concept is a three second neutral response that was developed by marine mammal trainers. The objective is to not reward a behavior that has just occurred. The theory is based on research that details that a reward or punisher that occurs after this 3 second delay is no longer rewarding or punishing the behavior that occurred. For me, the LRS is more than a technical tool. It is an opportunity to recollect my thoughts and move ahead more appropriately rather than becoming reactive. I use it on people all the time. My training team laugh about me looking through them for those three seconds, saying things like 'Oh boy, I am in the dog box, she is not responding.' It is my opportunity to take a deep breath and become present once again. All the gurus have said this for centuries. Come back to your breath. Breathe. This is a tool that we use that helps us to stay conscious. A true self awareness lesson.

Accountability and Responsibility: In order to progress, an alternative to being a 'control freak' is to empower ourselves and animals with the ability to respond. Focusing on 'what is' and what we can do to make a difference rather than on the overwhelming aspects of life that we cannot control.

Teaching us to Feel: The animals train us to feel. We will notice that we are more successful when we are intuitive. We cannot be intuitive unless we are present. To be present, we

Chapter 16 Ego – Friend or Foe

must be conscious of how we feel. Cowboys need to admit that they do cry after all.

Building Trust: I play a game in the animal training workshops that I offer. Many of you have probably played the game in various scenarios. It is the trust game, where one person in a pair must stand straight without moving his feet, and fall back into the second person's arms. For many people this is very difficult. As a human race we generally battle with trust. When I brief people on this game I tell them to teach their partners, without talking to fall back into their arms. At times, I will even blindfold the person who is required to fall. Most people go into the exercise without any thought. When they have all fallen around the room, I will choose the least trusting person to come and be my model. I will reiterate that my instruction to the group had been 'to teach' their person to fall back into their arms. I will then demonstrate how to teach that person to fall. I begin by standing really close to the person, so they can feel my presence. I gradually step further and further away, ever so slightly approximating myself away. Doing it this way, I am not only teaching the person that I am dependable, I am also learning where my limitations are. In some of the courses the least trusting person has been a very large man or woman. I am never sure if I can catch them. That hesitance will come through to them and thus I need to be sure too. From the animal's point of view, the lesson is – that they can trust us. We have their back. We earn their trust however, it is not something that we achieve automatically. In training lingo, this trust building exercise is clear communication, and in effect, we call it successive approximation. If we are fair and clear and operate at a pace that enables the animals to understand what we expect, we are trustworthy. To do this successfully we need to be present with the responses we receive and respond to ensure that we maintain our trusting relationship with the animal. Assumption in life is a lack of clear communication.

The Bottom Line

Humility is the key to being present, making us better animal trainers and more supportive engaging human beings.

Conclusion

Tail End

Being in relationship with an animal is potentially a life altering experience. I have met and shared relationships with the most amazing animals. Some of them, obviously, have passed on during my career. A predominant thought I have whenever I lose a friend, is the one we all have when someone close to us passes on. Did I make the most of the time I spent with that animal? Someone once told me that animals spend time with us so that we have an opportunity to learn more about ourselves. They spend time on this planet as a favour to us. However, it is said, that their short life spans, time spent in the physical form, because they are such clear light spiritual beings in essence, are a favour to them. They then return into the pure whole energy from where they came. To return their favour to us, we need to maximise the time they spend with us. And learn the heavenly lessons they are here to teach us.

The success of any relationship is based on the efficacy of the communication system between the two parties. This communication needs to be clear and simple. In relationships with human beings, we very often fail to keep communication simple. Consider an argument you have had with someone close to you. These are usually filled with justifications, defensive stories and dredging up the past. The arguments rarely stick to the subject matter that we were trying to attend to when the debate began. We feel the need to be right rather than deal with the debate matter for what it is. In a philosophy course I once did, I was taught how to put arguments into mathematical equations. The first step in this exercise, however, was to see if the basis of the argument was valid. Most often, it is not. For example, 'you are bad with money and so our relationship is unhappy'. This is an illogical premise on which to base our argument. The relationship and our choices about it have nothing to do with our spouse's attitude to money. Putting the two concepts together is irrelevant. It would be logical and provide us with an ability to respond – 'responsibility' – if we were to put it another way. 'I feel you are bad with money <u>and</u> I am unhappy in the relationship.' This is no longer an argument, but has become a situation where we take responsibility for ourselves and our own feelings in the relationship. The starting point is our feelings on the matter. This is the point from which we can begin to resolve the debate. When we see it for what it is and take responsibility for the part we play in the relationship.

From this we see that the communication system cannot deny how we feel. Our feelings will affect our ability to communicate, whether we like it or not. For this reason we need to remain totally conscious of how we are feeling, and use the feelings rather than try and deny them. We are, and we are working with sensorial beings that need clarity in order to understand us. If we admit to the feelings, we will be clearer, because we will be conscious of what we are communicating.

This book is a record of my experiences, and an outline of my opinion of these experiences. I was motivated to write my thoughts down after watching two individuals play the trainer game. A game where one person pretends to be an

Conclusion Tail End

animal, and the other the trainer. Using no words, the trainer has to teach the trainee to do something. During this game, I noted that although everyone was focussed on the trainee in the game and how well the trainee was doing in the session, the problem for the trainee was actually the trainer. What I noted in that moment was that sure it is vital to go through life learning the lessons we can from life, however, in this, we can only control our way of looking at the lesson. We can never control the responses of the animals we train, or for that matter any other being with whom we interact. How others respond to our interactions with them is always their responsibility. We will try and assume that responsibility for our children, or those we love, or simply those we wish to control, however, they remain individuals and it is always there responsibility. If someone steals a bicycle from us, we can choose to feel like a victim to the circumstance, or we can choose simply to accept the fact that we are now simply a person without a bicycle. We are responsible for the way we feel. Nobody else makes us happy or sad. We choose that feeling. Let's get back to the trainer. As a trainer, I cannot control the responses of my subject; however, I can be very clear in my communication. As clear as humanely possible. Because then, I will be giving clear feedback. If I have anything in my state of being that is affecting my ability to be clear, then I am not being completely responsible in my ability to train.

If we are not being one hundred percent truthful to ourselves about how we are in relationship to the people around us, we can never be truthful to the people around us. This is not making a judgement about how we feel about those around us. It is simply a statement about telling your truth. Providing clear and honest feedback in every scenario to those around you is telling the truth. If someone does not know that their habit of complaining about their health is driving away their friends, how can they change? If we want to do the kindest thing in our relationship with that person, we need to provide them with respectful feedback so that they can choose whether they wish to adjust. And then we need to provide them with the space to choose.

To round off, this is exactly what our relationships with animals is based upon. Clear and honest respectful feedback that provides the animal with a choice as to how it wishes to proceed in relationship with us. Fulfilling relationships with animals is what I choose. To achieve this, I strive to be humble enough. When I am humble, then I can hear what they are saying, and I can laugh at myself. True humility does not support inaccuracies or deceptions. It does not pity. It does not presume or assume. In a state of humility we are unconcerned with the stories of the ego. We let go of our need to impress and enable ourselves to be in the moment. We are clear on what needs to be done because we are seeing things as they are. If we support each other without judgment to be honest and open in all aspects of our lives, we provide each other with the space to be clear and present. I have been fortunate to spend time teaching hardened criminals about animal training. Watching a rehabilitated murderer love an animal is testament to the fact that the manner in which people treat animals is a mirror of the way they treat humans. First and foremost in that rehabilitation process is a need for the individual to forgive themselves. Only then will they have sufficient self-love to love another.

Sadly unethical dominant animal training practices do succeed, but don't guarantee the holistic well-being of the animals in our care. To be in a respectful relationship with an animal enables far more to be accomplished. The word enthusiasm literally means 'the God within'. The feeling of enthusiasm is only possible when we are present. It is possible that when we are present, we are inspired by something greater than ourselves.

There is mystery in that which we don't understand. If we accept this and simply work within the greatness of possibility, we will be humble enough to understand that nothing needs to be understood."

Appendix

An Introduction to behaviour Modification Theory

Touching Souls
Animal Training Foundations.

Knowing the word is owning the tool.

Why Bother with the Lingo?

What we learn from experience is that the application of insightful gut instinct plays as much of a role in our successful training programmes as the successful application of theory. If we apply a technique without a discerning understanding of how that technique will affect the animal, it may not always work. The theory can be confusing, and if we try and go through the list of definitions and concepts looking for a solution, without the right amount of relevant experience and intuitive consideration, we will probably fail to apply the theory successfully. Furthermore, in order to diagnose problem behaviour in an animal with any amount of credibility, knowing the theory will always clarify why what is occurring is happening.

For example, there is the classic tale of an animal keeper who was very concerned about the fact that the polar bears in the exhibit she was managing were constantly displaying stereotypic pacing. She feared for their sanity, and these fears were cemented after the keeper attended an enrichment workshop where she learned that stereotypic pacing may be a bad behaviour. At the workshop she learned about the power of enrichment activities, and came away with a number of ideas on how to stop the pacing; enrichment ideas that would increase the interest that the polar bears had in their exhibit. One of the ideas was ice lollies filled with frozen fish. The keeper made some of these lollies, and the next time she noted the polar bears pacing, she tossed the lollies in their path. The pacing stopped and her managers were amazed and grateful they had sent her on the course. However, in a couple of hours, the bears were up to their old tricks. So, the keeper made more ice lollies, and in order to ensure the fish would last, put smaller pieces into the lollies, and in some lollies only put fish blood. The lollies were tossed into the enclosure on a few more occasions before the keeper noted that the pacing behaviour had actually become more severe, and furthermore, whenever the keeper was close by, the pacing behaviour would actually commence. She had effectively conditioned the bears to pace, by rewarding them for doing so. Her intentions were in the right place. She wished for nothing more than the well-being of the bears. But she did not have the necessary behavioural knowledge to enable her to effectively enrich the bears. Examples of inadvertent training occurring between pet owners and their charges are huge. Not only is a good knowledge of the theory of training required. An understanding of the natural behaviour of animals with which we interface, even domestic animals, is also vital. But we must never discount the feeling side of things.

Appendix An Introduction to behaviour Modification Theory

Nothing replaces a good combination of experience, gut feel, confidence and know how.

Recognising that the available psychological theory is not the be all and end all that contributes to our successful relationships with animals has resulted in my opinions, and these opinions are the foundation that has developed into this book. I have noted, that even with a thorough developed theoretical base, there are still some trainers who don't succeed. For a long time I thought that this was because of their inability to be intuitive. This consideration baffled me. I believe that we are all able to be intuitive, and when I investigated the concept of intuition, I found that it is inherent in each of us. There is something even deeper than intuition at play. The bottom line that drives our success in our animal behaviour modification relationships, are our belief systems and the feelings we have about ourselves. These determine our capacity to develop and utilise our intuition as well as our ability to communicate effectively. Certainly, a good thorough knowledge of the theory can make us more confident, and nothing can replace experience in this field. However, unless we recognise and internalise our abilities and skills, at the same time as truly know who we are at the core of our beings, we will never achieve even a small part of our great potential, or be completely fulfilled in relationship with animals. Furthermore, if our knowledge is not internalised, we will not be present in the training session, worrying too much about the outcome rather than what is happening in front of us. When teaching trainers, the most difficult lesson to teach students is timing. They must not think that something the animal is doing is right or wrong and then respond. They have to respond as the animal is doing the required behaviour. This timing is fundamental to effective communication. And it is only possible if the trainer does not hesitate. And not hesitating means that the trainer needs to be in the moment, doing what they can without worrying about the outcome.

To be clear in our communication with the animals and those around us requires a technique. We cannot put the horse before the cart unless we know how to attach it properly; and attaching it properly is the theory. A successful training programme requires a chosen philosophy, so that when we enter into communication with the animals, the language is clear. When speaking to English speaking people, the greatest measure of success in the conversation will be achieved if we speak English, not Russian, Italian or Chinese.

I consulted at the Prague zoo and had a wonderful lesson about clarity in communication with the animals. The trainers were working with South African fur seals. The training in the zoo, generally was excellent. I found that my job was simply to fine tune their skills. There was one behaviour, a medical lay out that they were having difficulty training. They asked if they could demonstrate and then I would comment. They brought the animal out and conducted the session in the native Czech language. I have no understanding of this language and even found the intonations confusing. So I was hearing the language from the animal's point of view. This made it very easy for me to communicate with the trainers that the problem was very simple. Their communication was not clear for the seal. I could not understand what the trainer wanted, and when the trainer said the word good to indicate that the animal had achieved something correct, so I was pretty sure that the animal was a little unclear too. Apparently the Czech language has a variety of sounds that are used to indicate different tenses, and these sounds make the word good sound different in different scenarios. The entire session I had not been able to figure out which word they were using to tell the animal that it had done the correct thing. If I could not figure this out, then there was a good possibility the animal was having a hard time working this out. They reverted to using a whistle as a way to indicate that the animal had delivered the required response, and in no time the seal made good progress.

There are so many exciting tools and theoretical definitions and practises out there. I remember attending a marine mammal training conference with a colleague of mine. My associate is one of the better trainers I have ever met. She is incredibly

Appendix An Introduction to behaviour Modification Theory

intuitive, and has phenomenal relationships with the animals she trains. She has a fantastic photograph of herself while she is diving with a dolphin and the dolphin who was in labour at the time, is lying cradled in her arms. The trainer has a great deal of experience working with animals, and basically, if she is given a task to train an animal to do something, the result is generally guaranteed from the outset.

My friend did not enjoy the conference. She did not understand the terminology that the international trainers were tossing around, and made no bones about telling them to simplify what they were talking about. Not being as brave as her, I simply sat and listened. Some of the terminology is very confusing. What I recognised however, was that in our discipline, to achieve success, whether we know the theory or not, we all use the defined techniques. For trainers that are not trained, they too use the techniques, but their application is not as refined because they are not choosing techniques with clarity. For example, many dog trainers teach their dogs the automatic sit to prevent the dog from jumping up onto you. This is easy to teach. Basically, the dog only ever receives praise and attention from the person if they sit. This results in dogs coming over to elicit attention and simply sitting next to you to achieve its goal. This common knowledge is dished out and is effectively trained for the most part.

However, I have seen instances where it is not applied with success because people have not understood the theory of the technique. In theory, if we teach the dog that in order to get attention from us, it must sit in front of us, when it jumps up on us, to ensure the success of the automatic sit to prevent the jumping, we can pay no attention to the jumping. The theoretical terminology for this is commonly referred to as DRI, or differential reinforcement of an incompatible behaviour. The complaint I have heard from people who did not find success with the automatic sit training, is something along the lines of 'he knows the automatic sit when he is calm, but when I get home from work, he cannot stop himself from jumping up on me.' You can imagine that in some way, that person has rewarded that dog for continuing to jump up at these exciting moments, probably by flailing arms and chastising loudly, thus ineffectively used the theory, which calls for the trainer to reward something that prevents the dog from jumping up as opposed to rewarding the dog for doing both behaviours.

I encourage trainers that I work with to learn the theory. This is because, as I develop in my career, I find that as I understand more and more of the terminology, I find more tools at my disposal that I can use to fulfil my animal training objectives and more easily understand and diagnose problem behaviour. It is as though the training becomes more refined. With experience, these tools begin to come naturally. There are many books, websites, list serves, organisations and articles out there that discuss the terms. I would urge any trainer to go out and learn as much as possible from a theoretical point of view. And keep going back to the terminology and remain abreast of new developing trends. What I have also found is that sometimes I think I understand a term, but when I read it say a year later, after possibly having had an opportunity to apply the technique, I have found that in its application it is very different from my initial understanding.

A young lady was a part of my training team at Sea World for a couple of years. In that time she was an apprentice trainer, and so required to learn the theory. Her contact with the animals was minimal, and she expressed enormous frustration at having to learn all the terminology before being allowed more animal contact. She ended up having to relocate and so was forced to resign. Fortunately her career as an animal trainer was not over. She managed to gain employment in a bird training facility, and her time with the animals increased. It was only a few months into her new job when she called us to thank us for the theoretical foundation she had been forced to develop in our workplace. She explained how it was enabling her to fast track her career relative to other trainers in her new facility.

Appendix An Introduction to behaviour Modification Theory

Training terminology and techniques are constantly developing and changing. This is a field filled with passionate individuals who are determined to do what is right by the animals they work with. This provides a dynamic and exciting atmosphere for debate and development. Just when we think we know it all, a challenging animal or creative animal trainer comes along and gives us a new lesson. In the time that I have been working in this field, massive shifts in thought have occurred. I watched a talk by the legendary Bob Bailey, a man who worked with the Brelands. The Brelands were students of B.F.Skinner. While watching that talk, I was astounded at how new the theory really is. I felt like I was watching history, but it was right there before me. Still developing if we are open enough to learn the lessons. For me the truth again - Humility is the key.

Operant Conditioning

Most of my experience training animals is through the use of operant conditioning using reinforcement. Operant conditioning is basically learning through trial and error. When working with animals, our objective is to stimulate a response from an animal. We term this response a behaviour. For instance, when asking a dog to sit, the behaviour is the dog sitting. When using operant conditioning in an animal training context, after the animal has presented the response, the trainer presents the outcome that will provide the animal with the experience that will either increase or decrease the possibility of a specific behaviour from occurring in the future. In the dog sitting example, if we wish a dog to sit, and it does, we then reward the dog when it sits. The favourable reaction that the dog experiences when it sits will increase the possibility of it choosing to sit the future.

Operant conditioning uses punishment and reinforcement

Reinforcement

The word reinforcement, in the construction business means – to make stronger and sturdier. In the training sense it means exactly the same. It refers to something offered to the animal by the trainer that makes the behaviour response stronger. We colloquially often refer to the reinforcement as a reward, but we can see that the word reward does not tell the whole story.

Primary Reinforcement

A reinforcer can be primary, which means it satisfies a biological need, so this would include food sex, shelter, and social companions – anything that the trainer does not have to teach the animal to accept a reward. My pig Cleo has, since she has been a piglet enjoyed her stomach being scratched. When we do this her eyes glaze over and she falls on her side, relishing the attention while the bristly hair on her back stands erect. This for her is potentially a primary reinforcer. For my cat however, scratching her stomach is definitely not a reward.

Secondary Reinforcement

This is a reward that achieves reinforcement value as a result of the animal learning to like it. For example, money is simply paper. It has not biological value for us. Yet, none of us will willingly tear it up, because we associate it with primary reinforcement. We can purchase things that satisfy biological needs with the money. Khethiwe (means 'the chosen one' in Zulu) was a really independent dolphin calf. She did not seem interested in being stroked, which was very unusual for a dolphin. We had to teach her to trust us enough to stroke her. This was accomplished by pairing the touching with food. Now she cannot get enough of our attention. The stroking has achieved secondary reinforcement value. She also finds us tickling her tongue to be a secondary reinforcer. It is paired with the primary of food, and so she finds it rewarding.

Punishment

Punishment is the opposite of reinforcement. It serves to make a behaviour weaker. We would use it, if we chose, to reduce a behaviour from occurring. There are other methods of reducing behaviour from occurring without punishment. And they may be more effective in the long term. Punishment is a quick fix, but does not address the fundamental concern. There are a number of reasons why punishment is the most difficult tool in the tool box to use properly.

First of all, there is a danger when using punishment that we teach an animal to tolerate abuse. When we look at the subject of abused women or men in relationship with an abuser, it may help us understand this statement. We often ask why an abused human stays in a relationship. In most cases, it is because the abuse did not start with great intensity at their first meeting. It may have begun with a gentle insult. Gradually it would have increased over time, all the while breaking down the individual's self esteem, and making them reliant on the abuser for attention. It is the same with animals.

If we don't get the correct intensity with the chosen punisher, say a smack correct the first time we smack the animal, and up the pressure step by step, we may be teaching them to be abused. We actually use this method of training and refer to it as counter conditioning. This is often used when teaching medical behaviours. We teach the dolphins to lie still with their tails on our laps so we can draw blood samples from their veins. This takes some time to train. We begin by only putting pressure in the area where the needle will go. This pressure is gradually increased and eventually changed for sharp pressure and flicking with our fingers or an elastic band. We eventually progress to a fine insulin needle and gradually progress to a thicker needle. This prick has to be maintained, and we will not take blood every time we ask for the blood position.

If the intensity of the punishment is too high the first time, we will have confused the animal and broken their trust in us. If trust is the foundation of the relationship, we have wrecked a fundamental.

If an animal is punished for a response, there is a chance it will perform that response again in the absence of the potential punisher, particularly if the motivation to perform that response is quite strong. We can never be sure what the motivation is for an animal to participate in an activity. For example a dog jumping up onto a table to get our attention may feel less threatened on the table than on the floor. This motivation to feel safer may be a very powerful motivating factor. Our level of punishment needs to match the offence, and yet, we cannot be sure without achieving mind reading what that level should be.

Besides the fact that we are unsure what the motivation is for an animal to participate, we are also dealing with a scenario where we are having to fix a behavioural situation where the behaviour has already been rehearsed. There are some harsh examples of this. I chatted to a lion trainer who works lions for movies. He told me that if a lion who has been trained to jump up on a person ever manages to knock that person over, he will never be used as an actor in open contact again. This is because the lion has had an opportunity to rehearse the first step of the kill. And once this is in his behavioural repertoire, he will probably do it again. Science tells us that the first thing we learn is always the strongest impulse, so to extinguish behaviour is very difficult.

For punishment to be effective, it has to occur ever single time the behaviour presents itself. If not, we start from scratch. To illustrate this, consider why an addicted gambler stands in front of a slot machine and gambles away his fortune. It is the excitement of the possibility. The jackpot phenomena. This is well illustrated if you research more on

Appendix An Introduction to behaviour Modification Theory

the schedules of reinforcement. Basically it means that a success will set back your training.

Punishment is only effective if it is presented at the exact moment of the animal's transgression. From a behavioural perspective, animals and human beings only have a three second memory. When playing the training game with human beings I often see this played out in front of me. This game requires that one person teaches another person a predetermined 'trick'. There are many different varieties of this game. The one I refer to here is when the trainer teaches by using a whistle or clicker to tell the animal person when they are getting closer to the goal behaviour. The animal person will present a behaviour and then literally wait three seconds. If no response comes from the trainer in that period, they will move on to present the next behaviour in an attempt to elicit a reward. With regard to punishment, if it does not occur exactly at the right moment, we are punishing the incorrect behaviour. Humans so often do this with their pet dogs. The dog is chasing a car, and the person cannot punish the animal at that point because they cannot get near the animal, so the spank it upon its return, effectively punishing it for returning and not at all for chasing the car. The classic is that the person will tell you that the dog knows because it is cowering as it approaches the aggressive posture and angry tone of the owner. The people neglect to recognise that the dog is simply responding to the picture before them.

To sum up. It is not easy to use punishment, yet it is important to know what punishment is, as it does occur, even if we don't intend it to happen. If a tree falls over as we are riding our horse and they get a fright and jump sideways, they may have an association about a particular area as a result of that area being punished by the falling tree. We need to understand this so we can work to rectify the problem. Punishment damages relationship that is based on trust and choice. Very often our motivation to use it is reactive. This reactive mind is egotistically based. 'How dare you do that to me?!' 'My way or the highway.' 'Don't challenge me, I am the boss.' This power play mentality is not training.

Utilising rewards rather than punishment will generate a foundation premise in relationship with the animal. This premise is that the animals that we train are motivated by our presence. In order to ensure this, we need to consciously ensure that we remain an inspiring presence in their lives. In essence, theoretically speaking, I train so that I become a secondary reinforcer to the animal.

The other technical reason why punishment is not a chosen method of training is that it is not as effective in the long term. This is because the animal is operating out of fear when it does not do a particular behaviour. If the fear factor is removed and the motivation to do the unwanted behaviour is still strong, it will likely perform the behaviour again. It is not a matter of the animal choosing for itself, it is only reacting to its environment. So, in the long term, the extinguished behaviour is not trained away, only avoided.

Trainers as Secondary Reinforcers

Secondary reinforcers are rewards that become positive because they are associated with something primary. We use the animated secondary reinforcer of clapping our hands with the dolphins. Obviously the reward value of the hands being clapped would lose its effect over time if we did not maintain the association of the hand clapping and food from time to time. The secondary reinforcer status also works for us

as reinforcers in the animal's lives. If we always feed an animal, we will become through association, a motivating factor in that animal's life. My cats recognise and interact with me more than they do with my children, and this is probably because I am the hand that feeds them.

Appendix An Introduction to behaviour Modification Theory

They associate me with the person who fulfils one of their primary needs.

The creation of secondary reinforcers is the classical conditioning theory that was researched by a scientist called Pavlov. An experiment that most psychology majors are familiar with is when Pavlov worked with dogs, ringing a bell every time the dogs were fed. Eventually, through association of the food with the bell, the bell ringing alone stimulated the dog to salivate. Further research by Pavlov's scientists revealed higher order learning, where the bell was associated with another unconditioned reinforcer with success.

Association is a powerful tool. If we look at the advertising industry we see it rife in our media. Selling motor cars using pretty women as the catalyst to inspire us to buy those cars, or deodorant, where our use of it becomes a status symbol, are common examples of how advertising agencies increase their sales using the theory of classical conditioning. Sadly, the unconscious responses that this type of marketing creates are also responsible for a great deal of self-destructive behaviour in the human race. An example is how young ladies believe they have to be thin to be beautiful, and how this has given rise to dieting disorders.

In our relationship with animals it is vital to remember that negative associations can also be conditioned. Our family pet, Sasha was rescued as a puppy. She had a terrible illness, and we were even advised to have her put to sleep. We refused, and a vigilant treatment plan was put into place to nurse her back to health. Part of the treatment required many vet visits. Eventually Sasha would shake with fear every time we drove into the vet parking lot. She associated the visits with painful procedures. To alleviate her anxiety, we began taking her to the vet for pleasure visits. A few of her favourite Vienna sausages and some hugs and cuddles from the vet nurses were a part of these visits. We had to change the association to get rid of the anxiety.

Touching Animal Souls

A secondary reinforcer that can be used with horses is us gently stroking them. Remember that horses actually don't like being patted, so avoid this. A stroke is gentle and more acceptable to them. A pat is pressure, and can actually be used gently to get them to move. Gandalph and Mushatu are our two horses at home. Mushatu is a big thoroughbred, and he easily herds Gandalph out of a space by simply using his body language. We use the same technique when training the horses. Using simple body language to communicate to the horses to move in a particular manner. When Gandalph has moved away, Mushatu will stop herding him and both horses will comfortably go back to grazing side by side. When we have moved a horse in a training session using our body language pressure, we will change our body language to take that pressure off the horse at the exact moment of the horse complying. And the horse will visibly soften once more. As our relationship advances with the horse, we will begin to stroke the horse when we take off the pressure, and are in close proximity to the animal. Thus pairing the stroke with something that the horse finds inherently motivating – the reduction of pressure and the release from activity. It is not long before the horse will be responding to the stroke as though it is a reward; a secondary reinforcer.

I worked with a Nooitgedacht pony that had never been touched by a human before. We had to herd it into the round pen so we could begin the lesson of communication. Within half an hour, I had the horse leaning on my arm as I gently stroked its neck as a reward for it letting me yield its head to the side. When the lesson began he was avoiding me at all costs, and visibly shaking in fear. Without any undue pressure, the pony, minutes after we began working in the round pen, had followed me around. As he lay with his head in my arm I could not stop tears falling down my cheeks as I considered all the horses who are 'broken'. To break in a horse is still a term used to define if it is rideable. A well chosen term because using traditional horse training methods, their spirit literally has to be broken if they are handled in a manner where they are forced to submit. This

Appendix An Introduction to behaviour Modification Theory

is generally done by forcing a saddle onto their back and then having an able person sit on them till they stop bucking.

I felt sad as I imagined that this horse that had begun the lesson so frightened could have, if it had been trained in the traditional manner, been subject to that type of method, which would have created enormous anxiety and fear in its little soul. I was overcome with gratitude at the lesson Gandalph gave me. The lesson that allowed me to enter into this trusting relationship where I was able to generate a secondary of such vast and incredible worth within half an hour of meeting a horse. Where my touch would be a comfort to a frightened pony who only a half hour before had seen me as a predator. I was crying in gratitude, but also in sadness for what we as a human race do to animals in our desperate and at times even well intentioned efforts to control and 'take care' of them.

To maintain ourselves as a positive association for the animals is a task that requires conscious consideration. Punishment would not uphold this association. Furthermore, if we train animals in a haphazard fashion, and are inconsistent in our approach, there is a strong possibility that we will confuse and frustrate them with training methodology that is not clear. As a result, we would not be an inspiring and positive presence.

Motivation and Natural Behaviour

In view of the fact that we need to maintain ourselves as a positive force in the lives of the animals we train, it is important to remember that different species and individual animals are motivated by different things. For example, when approaching a horse, flowing soft movements are required. If we are jerky and frenetic, there is a strong chance we will unnerve the horse. On the other hand, when working with dogs or dolphins, they are often more motivated when we are more animated. The natural predisposition of animals, clear productive communication and consistent techniques all provide tools that will enhance our disposition as a beneficial presence in the animal's life.

Touching Animal Souls

What various species and individuals are motivated by is also an important variable when determining what and how we will train an animal and how we will reward it. Most domestic pigs, for instance are not high energy animals. If we ask them to run around as a reward for behaviour, they will not be motivated to participate. However, some pigs, like Cleo love their bellies to be scratched. Dolphins on the other hand are motivated to do high jumps and fast swims, and these behaviours can at times be used as satisfying rewards. Horses find that reduction of pressure rewarding. So do many herbivorous and prey animals. Rhinos, antelope and elephants to name a few. Their motivation is to do very little. Thus it is a very effective reward. To let it go back to doing nothing but graze or browse or simply stand still. The first thing anyone needs to do when entering into a relationship with an animal is learn what their natural behaviour and capacity is. This information will make it much easier how to work with the animal. It will, for example not be easy to teach a tortoise to move at speed into an enclosure or turn around really quickly. It may however be possible to teach it to gently plod along in a slow march in a requested direction. We also need to recognise that individual animals have different motivations. These are affected in part by their social standing, experience and individual demeanour. I worked with a couple of South African fur seals called Illanga (Sunshine in Zulu) and Moya (Wind/air/spirit in Zulu). They were both young pups when they were rescued and brought to our facility. They were both in a similar condition on arrival and roughly the same age. To teach Illanga to eat from us was no problem. In fact, we simply walked up to here, while she was sunning herself and lying down comfortably and waved a fish in front of her face, which she gently took and ate with gusto. Moya on the other hand bit many of us in the process of training her to eat, which had to be conducted, initially, as a force-feeding exercise. The process has been elaborated in the book. Their personality types remain different. Both are smart animals, but Moya is a skittish while Illanga is very calm and relaxed. Their predisposition affects the

Appendix An Introduction to behaviour Modification Theory

manner in which we work with them. With Illanga we are naturally confident and relaxed. Because Moya is less predictable, we are more conscious when in training sessions with her. To avoid being bitten and to ensure we don't teach her to be aggressive. We always need to be balancing our own energy and expectations to the particular personalities of individual animals.

Bridge

We have coined a term in animal training called the bridge. It is the way we tell the animal they have done what we have asked them to. In human terms, it is the same as someone congratulating you for an action, and you knowing that the compliment promises good things to come. If we wish to tell an animal well done, we 'bridge' for a behaviour. We develop a method of communicating with an animal ensuring that the animal understands the communication to mean – good job. Some commonly used forms of communication are to blow a whistle, click a clicker, tell an animal good, or similar. The word is 'bridge' because it connects the action we wish to reward and the reward. It is a quick effective symbol that the animal notes at the point where they have done the required behaviour. We offer the symbol. Thus, if we wish the dog to paw a ball on the ground, the moment his paw hits the ball we will present the bridge, which will be his cue to stop and receive his reward from us. These tools do not mean anything to that animal until the communication has been taught. For example, if I took a clicker and went and clicked it in front of a horse that had never heard the clicker before, it would not mean anything. It is the equivalent of someone telling you well done in a language that you do not understand. The bridge must be taught by being paired with something positive. In effect, in the Pavlov scheme of things, the bridge eventually becomes a secondary reinforcer.

The second important aspect of the bridge is that in order for it to be an effective form of communication, it must occur at the exact moment that the animal performs the behaviour that is required. We want the bridge to be able to pinpoint the exact moment in time that the animal performs the correct behaviour. When used properly, animals catch on to the meaning of the bridge very quickly. When used ineffectively, what will more than likely occur is that the animal will respond to the bridge, if it is rewarded often enough after it hears the compliment, however, there will be no progress in the training. So, timing is key. For example, if we want a dolphin to touch its tail onto a ball, we will blow the whistle as his tail touches the ball. If we blow it after he has touched the ball, he will be hearing the message that he is being rewarded for his tail being close to the ball. Not for touching it. Chances are, he will miss it second time around, because that is what you have asked him to do.

When working a polar bear through an acrylic window, my friend Franta, the lead trainer from the Prague zoo used a torch light as a bridge. He would flash it at the bear to indicate that the animal had delivered the required response. He did this in a very effective exercise. He was working with a polar bear that was initially exhibiting seven hours a day of stereotypic swaying. The animal was on a breeding loan. The trainer, Franta began by working with the animal at the exhibit window. He taught him a couple of behaviours through the window, all the while using the torch to highlight the exact moment of triumph which heralded a fish being tossed over the acrylic window to the bear.

What was remarkable about this training was that it took only about a month to reduce the bear's stereotypic behaviour to only three hours a day. It may be that the stereotypic behaviour was as a result of the bear being fairly anxious in his environment, and this is understandable when we consider that a large acrylic window at the interface of the exhibit is one that was completely out of the bear's control. This interface had a changing face with public coming up to it all the time. Bears traditionally spend a lot of their time exercising, and the

Appendix An Introduction to behaviour Modification Theory

path past the acrylic is in the water zone, so the animal had no real choice about avoiding the area or not. With the area becoming a positive one, due to Franta's training, it is possible that this contributed to alleviating the bear's concerns.

We like to think that the philosophies that we present in this book will help to bridge the gap between ourselves and those with whom we are in relationship by improving clarity of communication. Clarity in relationship always generates a more harmonious lifestyle. The bridge is only used to effectively improve our timing. I have noted time and again that the animal seems to know it has done the right thing an instant before the bridge is used. I believe that is because they are so tuned into our body language, and even while watching the most experienced trainers, I have noted a small softening in their body language an instant before they sound the bridge. With horses many people do clicker train them with success. I have tried this technique but find it can be difficult to apply effectively.

Horses are big animals, and a dominant horse will determine where it wants to eat. A submissive horse will give up its grazing position to a dominant horse. When using food as a reward for horses, and clicker training them, we need to be careful that we don't create a relationship with them where they begin to mug us for food. This requires excellent timing, and very often, a firm hand. I remember watching a very experienced seal trainer try her hand at clicker training a Shetland pony. The seal trainer was making good headway teaching the horse to target to her boot until the horse worked out that the food was in a pouch at her waist. He pushed his head into her waist and she lost her footing, and nearly fell. The horse remained determined and came toward her pouch once more. The trainer had some pellets in her other hand, and in an attempt to persuade him to move his head in another direction, coaxed him away from her middle with the food, effectively rewarding him for going after the food in her pouch. This was just a little pony.

Touching Animal Souls

When using negative reinforcement with horses, we actually don't have to use a bridge. If our timing is perfect and we take the pressure off at the exact moment of the horse doing the correct behaviour, the reward is us taking off the pressure. I have however had success with a vocal bridge and even a clicker indicating that I am about to take off the pressure. When working with trainers who battle with moving their bodies to take off the pressure at the exact moment I have actually let them use a clicker and then this almost cues them to take the pressure off directly after they click. This has been a good tool to use to teach people more effective timing, and it has actually assisted horses, who quickly understand this – more quickly, very often, than the trainers.

While helping to teach horse trainers to use the release of pressure as a reward, I found that one of the trainers who was having difficulty getting close to his horse finally make a breakthrough. From a distance I saw him standing touching his horses nose. I was about to congratulate him for his efforts when I saw that the horse was chewing pellets that had just been fed to him. I breathed deeply because the trainers had been told not to use food. I asked the trainer to leave the round pen and go and put his food away. I then asked him to get back into the pen and ask the horse to come towards him. The horse came up to the trainer immediately and then realised there was no food, so trotted off to resume his distance. He had not learned anything from the pellet bribery. This time round there was no food, so no motivation to be with the trainer. My interpretation of this was that the horse had simply found food the first time round. He did not make a conscious choice to get closer to the trainer. He had not learned anything in relationship with the trainer.

Appendix An Introduction to behaviour Modification Theory

I have found that horses are more focussed and alert when being taught using pressure and release. Thus, even if we are using food, we need to ensure that the posture we are using to direct the animals remains a primary focus in ourselves. In a yoga class recently I realised how little consciousness I have about what my physical body is up to. Three times a week I attend the class and it is in those moments, when my instructor tells me to sit quietly and contemplate a movement I have just done that I achieve a body consciousness about what I am feeling. I realised that this is the level of consciousness required for me to be completely effective when working with animals, particularly when using the pressure and release method.

When training in this manner, we do not use a bridge, but the horses are extremely tuned into our body language, and the moment we soften, they will see this as the reduction in pressure that is naturally motivating for them, and so, the inherent reward. This is not to say that it is not possible to clicker train a horse. It is definitely possible. I have achieved success myself with the technique. It simply requires great presence of mind and a formidable grasp of the theory as well as a confident disposition in order to achieve and maintain success.

It also requires that the relationship of leadership is firmly in place. There are often difficult behaviours that I find easy to train when using the clicker. Gandalph needs sunblock on his nose every day because of his fair complexion. I did not realise how sensitive he was going to be to this application. The first time I put it on him he took off galloping across the paddock. The smell must have been quite something for him. The next day he was not having any of it. When I approached him with the clicker and taught him to accept it this way, with carrots as a reward for allowing a gentle application of the formulation, he stood for the entire time and participated, sans halter, without any problem. I have an accomplished horseman friend who clicker trains his horses for carriage riding amongst a variety of other behaviours. This gentleman is successful because he maintains his role as master in the relationship with his horses. He does this very effectively. I have watched him work with a quarter horse cross and a Shetland pony in the same session. Both horses were literally playing football with him. It was a wonderful relationship in action. I have also recommended clicker training to people on occasion. For example, the police do work with their animals in a crush. In this scenario the animal cannot mug the less accomplished trainers for food, and the training will remain positive and progressive.

Most important with the use of clicker training and horses is not to lose yourself in the bridge. This literally means forgetting all the posture we are presenting because we are so focussed on achieving our target behaviour and clicking that clicker. If we lose our body consciousness we are more than likely going to have difficulty maintaining a sensitive alert horse.

Cues

The bridge is one part of the language that we use to teach the animal. It is the simple part, because the animal will soon associate it with good things, and respond with enthusiasm as a result. There are other parts in the language we develop with them. When working with animals using reinforcement, we can request particular behaviours on purpose by placing these behaviours on cue. Cues are signals, a stimulus that is our way of requesting a particular behaviour. We teach the animal the cue, and in effect, it becomes the language we share with them. Just like the word butter is not the butter, but simply a name for the butter, so too is a cue a symbol for a particular behaviour. For example, when we want a dog to sit, we say sit. The dog does not understand the word sit, but associates the sound of it with the activity that has been associated with the sound. He knows that when he sits after hearing the sound he will receive affirmation.

Appendix An Introduction to behaviour Modification Theory

Dolphins hear in enormous detail. In fact their hearing is probably their most important sense. They hear almost ten times better than we do with a much larger frequency range. For this reason, we don't generally cue them using words, but choose hand and body signals to cue behaviour. The association is done the same way when we train. For instance, when we clap our hands, the dolphins can be taught to associate seeing that cue with slapping their tails on the water surface.

Horses are generally taught to yield to pressure. They are incredibly sensitive to this touch. A horse can be taught to halt when we ride it by moving our body weight backwards. A slight leg pressure on a particular part of their body can cue a part of their body to move. In this sense, the cue is a tactile signal. And all these cues should be taught as part of the initial ground work, before we even get into the saddle. A sensitive horse is a very achievable task. If we apply the techniques of training consistently, we can ride a horse on a bitless system and conduct dressage manoeuvres. I have seen this achieved with many different species and temperaments of horses. Horses also respond very well to verbal cues. My carriage driving friend directs his horses with vocal cues. Posturing cues are also very effective with them.

What is interesting about my horse riding lessons all those years ago is that I was taught to ride a horse like I was taught to drive a car. Kicking is the accelerator, yanking one way or another to make the turn like the steering wheel and pulling to put on the brakes. I was taught this as though it is a universal language. Since I have learned to train horses I have noted that any cue that the horse is taught is the one he will respond to. My carriage horse friend uses vocal cues to drive his horses. The horses respond with finesse and grace. Horses taught on a bitless system are not responding to mouth pressure. They need to be taught to respond to different feelings. To be ethical in relationship not only with horses, but with all animals, we need to ensure that we are clear with our cues, and that we are sure that the animal is responding to the cue we have taught. Our clarity will be their peace of mind. It will generate confident participation.

Difference between Positive and Negative Reinforcement

The confusion that these two terms cause in the animal training world is a classic example of the reason for the fall of mankind. Positive simply means adding. Negative means subtracting. Because of human greed, we have attributed the word good to positive and the word bad to negative. We need to forget our human fear of loss and lack when listening to the language of training. Positive and negative can both be reinforcement techniques. Positive reinforcement means adding something in order to make behaviour stronger. For example, providing food to a dolphin as a reward for it accomplishing something in order to ensure that the dolphin does the behaviour again. Negative reinforcement is removing something that will serve to make behaviour stronger.

With horse training, this may mean the removal of pressure. My Prague Zoo friend contacted me when he and the keeper of the horses was asked to assist with the training of a Przwalski stallion. The horse had never been willingly touched by humans before. It was on its own in a camp because of breeding controls. The animal was unmanageable and the keepers needed him to be more manageable and allow them to gate him from one paddock to another with ease. Franta emailed me and asked for my assistance. I emailed a reply, explaining to him how to basically put gentle pressure on the horse if it was moving away from him, and then when the animal turned to face him, to back off, taking away his pressure and turning his chest inwards away from the horse, thus negatively rewarding the animal for his attention. Franta was amazed at how quickly the technique worked. He sent me a video of the proceedings.

Appendix An Introduction to behaviour Modification Theory

It is clear that this man has great sensitivity and timing, which is the reason for his success. It was a delight to watch this stallion and keeper getting to know each other. In a few sessions the animal was following Franta around like a puppy. The horse had a confident disposition and eager willing look in its being. All due to the correct use of negative reinforcement.

Another example is if I want a horse to move backwards, when I push him I must stop pushing the moment he moves. He understands that the pressure will reduce when he is doing the right thing, and thus knows he had done the right thing when the pressure is removed. This is negative reinforcement in action. The pressure, or influence comes first, and the reinforcement is the removal of the pressure. The word negative refers to the removal of the pressure. An easier to word to use in this instance is subtraction reinforcement. The easiest example to understand is when we are in a car and the beeping noise is only alleviated when we put on our safety belt – the car negatively reinforces us for putting on our safety belt.

Classically negative reinforcement was thought to have to follow an aversive stimulus. I think it is important that we are clear that this is very much an anthropomorphic description of events. For example, with horses, it is natural for them to use body language and pressure to communicate with each other. So, if we use this method to train behaviour, it could be interpreted very differently from the horse's point of view. Furthermore, if we use food as a positive reinforcer, are we not utilising negative reinforcement too. Because, when we provide food as a reward, we are adding something to the environment that may be taking away a feeling of hunger. Basically it is important that we are always aware that the linguistic terminology that we use to describe training may be very differently interpreted by the animals that we are training. However, whatever method we employ must always be used with sensitivity and to suit the needs and personality of the particular animal with which we are interfacing.

Horses are incredibly sensitive animals, and when we work with them properly, the amount of pressure that we have to use is so slight, it is simply a cue. If we are consistent with our communication process, the horse will remain responsive and comfortable. In fact horses that are worked consistently and clearly in this technology seem to be in a productive working relationship with what appears to be their leader, or rider. Pressure is not necessarily physical. The key is the manner in which trainers will approach a horse to begin a session. If the horse is running off, the trainer will continue a determined follow. The follow is the pressure. The moment the horse stops and looks in the trainers direction is the moment the trainer stops and backs off a little. I have seen this method work with the most skittish and difficult horses, to the point where the horse is eventually walking up to the trainer to elicit the session.

The word positive in the term positive reinforcement refers to the addition of something which serves to reinforce behaviour. For example, when the dolphin responds favourably, we feed a fish or provide a stroke, or an ice block. All adding something to strengthen a desired behaviour. The fish arrives after the completion of the behaviour.

Both negative and positive reinforcement serve to increase the probability of a behaviour occurring in the future.

Schedules of Reinforcement

In terms of motivation, and to aid in maintaining ourselves as motivating forces in the lives of the animals we train, there are other theoretically proven measures that we can adopt that enhance this possibility. The jackpot phenomenon, for example is detailed in the literature as a reinforcement schedule. The trainer carefully chooses when and how to reward behaviour, and in this manner can maintain high motivation levels

Appendix An Introduction to behaviour Modification Theory

in the animal they are training. A fair amount of research has been conducted into the subject. In the training programmes I am familiar with, we have employed a variable schedule of reinforcement using a variety of reinforcers. For example, with the dolphins, we use a variety of secondary reinforcers, such as strokes and tickles, their favourite high energy behaviours, toys, and our animated attention. These are used in conjunction with food.

So, when the animal does something and is about to be rewarded, they never know what the reward will be, or if there will be a reward at all. When employed properly, this system yields wonderful results and an excitement and anticipation in the animal that holds the potential of a greater success than if we were to employ a more predictable schedule of rewarding the same reward every time the animal responded appropriately. Secondary reinforcers, as already discussed are powerful if used properly. Obviously, in order to achieve the greatest amount of success when using the variable schedules of reinforcement requires an experienced, intuitive and confident trainer who has an established relationship and an informed knowledge of the animal they are training. Furthermore, it is important to bear in mind what is individually and naturally motivating for the animal you are training. Some animals require more consistency in the type of rewards they receive, and this could influence the reinforcement schedule that you choose to use when training.

So How do we Train?

I was new to the world of horse training, and a friend of mine, Dave was holding a 'clicker training' workshop with horses, for some folk who were very green to the concept of operant conditioning. I had been using the technology for years with other animals, and had made friends with this horse training instructor because I was interested in the application of this type of technology to the horse training world. The audience on that day were made up of a bunch of people from varying backgrounds. Some were keen equestrian professionals, a couple were horse racing enthusiasts, and there were even some novices who were experiencing problem with their horses and looking for novel solutions and general knowledge on the subject. There was a fair amount of excitement as they watched the practical aspects of the training that Dave was demonstrating on one of his more seasoned ponies called Phillipe. He had trained Phillipe to do some basic riding behaviours such as lunging and responding to leg pressure using the clicker. He had also trained the pony to do some 'tricks', and the most endearing had to be the horse mimicking his trainer by crossing his front legs if the trainer offered the cue of crossing his own legs.

The oohs and ahs of the people watching showed they were impressed. I too was impressed, and applauded with the rest. Then a more sceptical lady asked the big question. 'This is all very well with a seasoned horse, but how do you get a horse to understand the concept'. My friend smiled. Everyone there had been told earlier about a new pony that had arrived a few days before. The animal had been schooled in the traditional horse training manner, but had never been clicker trained. The horse had been acquired to become a part of the horse riding school that made up a part of my friend's business. Dave turned to me and asked me to fetch the pony, whose name was Silver, and clicker train it in front of everyone. As I left to fetch Silver he explained to the audience that I was familiar with the manner of training, but had to date never used it on horses. This was true. He also reminded the audience that Silver had never been clicker trained in his life.

I brought Silver into the arena. The pony was calm and focussed. I was a little apprehensive, but excited about the prospect of exploring the possibilities. I clicked the clicker and fed the pony a few cubes and spoke gently to him. His ears pricked up and he was alert. I noted his enthusiasm and

Appendix An Introduction to behaviour Modification Theory

decided to waste no time. I touched a rubber ball that I held in my hand to his nose, clicked and repeated the reward process. Silver's eyes sparkled. I repeated this step, but this time let go of his halter at the same time. He did not run off, but stayed focussed on the task. Then came the crunch moment. The audience was completely quiet. Dave was watching from the side lines with a confident smile on his face. He nodded to me gently. I held the rubber ball twenty centimetres in front of Silver's face. Silver looked at me quizzically for a couple of seconds, and then very deliberately raised his snout and touched the ball. My heart raced excitedly, at express train pace, I clicked and treated. I heard a disbelieving voice in the audience mutter 'no way'. 'Yes way' I thought. I had seen this before with other types of animals. I was so excited. I knew that Silver and I were about to demonstrate the profound communication that can occur in this type of training. I took a step back, stood a couple of metres away from the pony and held up my hand holding the ball. He would have to step forward to reach me. His quizzical expression had been replaced by a focussed concentration. He hesitated still, a little longer and then took the step, lifted his snout and targeted to the ball. The audience applauded with excitement, convinced at last.

This pony had never been trained with this method. Yet, the rules of this game are so simple, that they are easy for anyone to apply. What was amazing to watch, was also the fact that the pony was completely focussed on the task at hand, and even appeared curious. No coercion was necessary to achieve the result. My task was simply to ask clearly, and respond timeously and appropriately. Utilising negative reinforcement is also extremely effective. When used with horses, the method can be initiated with seasoned or green horses. In fact, if a horse is green, it is easier to train using this method because its natural response to pressure is still to yield. With horses that have been trained using traditional methods, they have often been desensitised to human pressure because it has been ineffectively used in their conditioning. How do we train using negative reinforcement with horses? Simple. We apply pressure, in the gentlest manner possible to illicit a response, and when the horse does what we wish it to do, we take the pressure away. So, if we wish the horse to lower its head, we place our hand behind his poll and gently ask him to lower his head with pressure. The milli-second he submits, even slightly, we stop asking. Then he knows what we are asking for.

This is the basic philosophy, and if we apply it consistently, we will succeed at training animals. However, to achieve a successful relationship with animals requires that we are successful communicators. This requires the use of what is broadly termed gut instinct. This book has investigated how what occurs within us affects our relationship with animals. The theory on its own will not sufficiently ensure our success. That is, of course, if we are working to ensure that fulfilling long term relationship with the animals. In my mind – why else would we choose to train them.

Bibliography

Blink (Malcolm Gladwell) – Penguin Books. 2005

Anna Wise – The High Performance Mind; Mastering brainwaves for Insight, Healing and Creativity. Tarcher/Putnam. 1996

Pets and Our Mental Health: the Why, the What and the How. (Johannes Odendaal) Vantage press, Inc, New York. 2002

The Power of Now (Eckhart Tolle) Hodder and Stoughton – 1999

Universal Principle Cards (Arnold Patent, Stephen and Kathleen Norval)

About the author

Gabrielle lives in Drummond, KwaZulu Natal on a plot that she shares with a menagerie of animal souls, her children, Zac and Kai and husband Darryl. She has worked as a professional animal trainer since 1990, with the dolphins, seals and penguins (and sometimes fish) at Sea World, and is also a trainer with Horse Gentlers International. She has consulted at a number of facilities, and presents animal training workshops. One of her favourite past times apart from being with animals is listening and appreciating her son's rocking musical talents.

If you enjoyed this book (and naturally we hope that you did) we recommend the following titles for your further reading enjoyment under our imprint.

Hunting with the Heart – A Vision Quest for Spiritual Emergence by Graham S. Saayman Ph.D.

After many years of studying large brained social mammals such as dolphins, baboons and elephants in their natural environment and developing a strong bond and respect for nature, emeritus Professor of Jungian Psychology Graham S. Saayman tells the reader of his psychic experiences whilst out in the field. He writes in a way that can only inspire the reader to desire the same bond with the natural world that Saayman himself has developed and through intimately personal and touching accounts of his experiences, the reader is encouraged to see the world around them in a new and beautiful way.

ISBN: 978-0-9802561-1-6

Happy Ever After – Building Relationships that Flourish by Hanna Kok

In the West fewer and fewer married couples are able to keep their marriage from ending in divorce. What has changed so much in the last few decades to result in divorce becoming such a big problem? According to Hanna Kok, a marriage is based on two main elements, love and respect. In Happy Ever After it is argued that there is a lot of focus put on love, but not nearly enough attention given to the concept of mutual respect. Written in easy-to-read language and organised in practical sections, Kok shows the reader not only how to make their romantic relationships last, but also how to get the most out of relationships that you have with colleagues, children and friends.

ISBN: 978-0-9814278-3-6